*Lab Manual
to
Accompany*

# Machine Tool and Manufacturing Technology

Steve F. Krar
Mario Rapisarda
Albert F. Check

Delmar Publishers

*an International Thomson Publishing company* I(T)P®

Albany • Bonn • Boston • Cincinnati • Detroit • London • Madrid
Melbourne • Mexico City • New York • Pacific Grove • Paris • San Francisco
Singapore • Tokyo • Toronto • Washington

## Notice to the Reader

Publisher does not warrant or guarantee any of the products described herein or perform any independent analysis in connection with any of the product information contained herein. Publisher does not assume, and expressly disclaims, any obligation to obtain and include information other than that provided to it by the manufacturer.

The reader is expressly warned to consider and adopt all safety precautions that might be indicated by the activities herein and to avoid all potential hazards. By following the instructions contained herein, the reader willingly assumes all risks in connection with such instructions.

The publisher makes no representations or warranties of any kind, including but not limited to, the warranties of fitness for particular purpose or merchantability, nor are any such representations implied with respect to the material set forth herein, and the publisher takes no responsibility with respect to such material. The publisher shall not be liable for any special, consequential, or exemplary damages resulting, in whole or in part, from the readers' use of, or reliance upon, this material.

**Online Services**

**Delmar Online**
To access a wide variety of Delmar products and services on the World Wide Web, point your browser to:
http://www.delmar.com/delmar.html
or email: info@delmar.com

**thomson.com**
To access International Thomson Publishing's home site for information on more than 34 publishers and 20,000 products, point your browser to:
http://www.thomson.com
or email: findit@kiosk.thomson.com

A service of I(T)P®

Copyright © 1998
By Delmar Publishers
A division of International Thomson Publishing

The ITP logo is a trademark under license

Printed in the United States of America

For more information, contact:

Delmar Publishers
3 Columbia Circle, Box 15015
Albany, New York 12212-5015

International Thomson Publishing
Berkshire House 168-173
High Holborn
London WC1V 7AA
England

Thomas Nelson Australia
102 Dodds Street
South Melbourne, 3205
Victoria, Australia

Nelson Canada
1120 Birchmont Road
Scarborough, Ontario
Canada, M1K 5G4

International Thomson Editores
Campos Eliseos 385, Piso 7
Col Polanco
11560 Mexico D F Mexico

International Thomson Publishing GmbH
Königswinterer Strasse 418
53227 Bonn
Germany

International Thomson Publishing Asia
221 Henderson Road
#05-10 Henderson Building
Singapore 0315

International Thomson Publishing—Japan
Hirakawacho Kyowa Building, 3F
2-2-1 Hirakawacho
Chiyoda-ku, Tokyo 102
Japan

All rights reserved. No part of this work covered by the copyright hereon may be reproduced or used in any form or by any means—graphic, electronic, or mechanical, including photocopying, recording, taping, or information storage and retrieval systems—without the written permission of the publisher.

2 3 4 5 6 7 8 9 10 XXX 02 01 00 99 98 97

**Library of Congress Card Number: 95–25386**

ISBN: 0–8273–7587–5

# CONTENTS

To the Instructor .................................................................................................................. 1

A Guide to the Use of this Workbook ................................................................................ 2

Test #1      Technical Drawings/Prints (Section 3, Unit 3) ................................................ 5

Test #2      Machining Procedures (Section 3, Unit 4) ..................................................... 7

Test #3      Accident Prevention (Section 4, Unit 5) ........................................................ 9

Test #4      Measurement Systems (Section 5, Unit 6) ................................................... 11

Test #5      Micrometers (Section 5, Unit 7) .................................................................. 13

Test #6      Vernier Calipers (Section 5, Unit 8) ............................................................. 15

Test #7      Inspection Tools (Section 5, Unit 9) ............................................................. 19

Test #8      Laying Out (Section 6, Unit 10) ................................................................... 21

Review Test #1      Measurement and Layout (Sections 5 and 6—Units 6 to 10) ...................... 25

Test #9      Manufacture of Iron and Steel (Section 7, Unit 11) .................................... 29

Test #10      Metals and Their Properties (Section 7, Unit 12 ) ....................................... 31

Test #11      Metal Cutting (Section 8, Unit 13 ) .............................................................. 33

Test #12      Cutting Fluids (Section 8, Unit 14 ) ............................................................. 35

Review Test #2      Metals and Principles of Metal Cutting (Sections 7 and 8—Units 11 to 14) ........... 39

Test #13      Hand Tools (Section 9, Unit 15) .................................................................. 43

Test #14      Threading and Reaming (Section 9, Unit 16) .............................................. 45

Review Test #3      Bench and Hand Tools (Section 9—Units 15 and 16) .................................. 47

Test #15      Cut-Off Saws (Section 10, Unit 17) ............................................................. 51

Test #i6      Contour-Cutting Bandsaw (Section 10, Unit 18) ......................................... 53

Test #17      Drill Press Types (Section 11, Unit 19) ....................................................... 55

Test #18      Twist Drills (Section 11, Unit 20) ................................................................ 59

Test #19      Producing and Finishing Holes (Section 11, Unit 21) ................................. 61

Review Test #4      Drill Press (Section 11—Units 17 to 21) ...................................................... 63

Contents      iii

| | | |
|---|---|---|
| Test #20 | Lathe Types and Construction (Section 12, Unit 22) | 67 |
| Test #21 | Lathe Accessories and Tooling (Section 12, Unit 23) | 69 |
| Test #22 | Cutting Speeds and Feeds (Section 12, Unit 24) | 73 |
| Test #23 | Mount And Remove Accessories (Section 12, Unit 25) | 75 |
| Test #24 | Mount Work Between Centers (Section 12, Unit 26) | 77 |
| Test #25 | Machining Between Centers (Section 12, Unit 27) | 79 |
| Test #26 | Knurling, Grooving, Shoulder Turning (Section 12, Unit 28) | 81 |
| Test #27 | Taper Turning (Section 12, Unit 29) | 83 |
| Test #28 | Machining in a Chuck (Section 12, Unit 30) | 85 |
| Test #29 | Threads and Thread Cutting (Section 12, Unit 31) | 87 |
| Review Test #5 | Lathes, Accessories, and Thread Cutting (Units 22 To 31) | 89 |
| Test #30 | Horizontal Milling Machines and Accessories (Section 13, Unit 32) | 93 |
| Test #31 | Milling Machine Setup (Section 13, Unit 33) | 95 |
| Test #32 | Horizontal Milling Operations (Section 13, Unit 34) | 97 |
| Test #33 | Vertical Milling Machines (Section 13, Unit 35) | 101 |
| Test #34 | Vertical Milling Machine Operations (Section 13, Unit 36) | 103 |
| Review Test #6 | Milling Machines (Section 13—Units 32 to 36) | 105 |
| Test #35 | Bench And Abrasive Belt Grinders (Section 14, Unit 37) | 109 |
| Test #36 | Surface Grinder Wheels and Operations (Section 14, Unit 38) | 111 |
| Review Test #7 | Grinding Operations (Section 14, Units 37 and 38) | 113 |
| Test #37 | Computer Numerical Control (Section 15, Unit 39) | 115 |
| Test #38 | How CNC Controls Machines (Section 15, Unit 40) | 117 |
| Test #39 | Preparing for Programming (Section 15, Unit 41) | 121 |
| Test #40 | Linear Programming (Section 15, Unit 42) | 123 |
| Test #41 | Circular Interpolation (Section 15, Unit 43) | 127 |
| Test #42 | Subroutines or Macros (Section 15, Unit 44) | 131 |
| Test #43 | CNC Machining Centers (Section 15, Unit 45) | 133 |
| Test #44 | Turning Centers (Section 15, Unit 46) | 137 |

| | | |
|---|---|---|
| Review Test #8 | CNC Operations (Section 15, Units 39 to 46) | 139 |
| Test #45 | Heat Treatment of Steel (Section 16, Unit 47) | 145 |
| Test #46 | Artificial Intelligence (Section 17, Unit 48) | 147 |
| Test #47 | Computer Manufacturing Technologies (Section 17, Unit 49) | 149 |
| Test #48 | Coordinate Measuring Systems (Section 17, Unit 50) | 153 |
| Test #49 | Electrical Discharge Machining (EDM) (Section 17, Unit 51) | 155 |
| Test #50 | Flexible Manufacturing Systems (Section 17, Unit 52) | 157 |
| Test #51 | Group Technology (Section 17, Unit 53) | 159 |
| Test #52 | Just-In-Time Manufacturing (Section 17, Unit 54) | 163 |
| Review Test #9 | Manufacturing Technologies (First Half) (Section 17, Units 48 to 54) | 165 |
| Test #53 | Lasers (Section 17, Unit 55) | 169 |
| Test #54 | Robotics (Section 17, Unit 56) | 171 |
| Test #55 | Statistical Process Control (Section 17, Unit 57) | 173 |
| Test #56 | Stereolithography (Section 17, Unit 58) | 177 |
| Test #57 | Superabrasive Technology (Section 17, Unit 59) | 179 |
| Test #58 | The World of Manufacturing (Section 17, Unit 60) | 181 |
| Review Test #10 | Manufacturing Technologies (Second Half) (Section 17, Units 55 to 60) | 183 |

# TO THE INSTRUCTOR

This workbook is intended to be used with *Machine Tool and Manufacturing Technology*. Its contents include a variety of tests which cover the subjects examined within the text. The questions fall into five different categories: completion, true or false, multiple choice, matching, and identification. Each test has been designed to review the material in the text thoroughly, and at the same time, provide a challenge to the students and to increase their knowledge of each topic.

In order to test the students fairly and for ease in marking, each answer has been weighed evenly and has the same value. This has been done so that the instructor can vary the value of questions at his or her discretion. The time required to complete each test has been left to the judgment of the instructor. As a rule of thumb, it should take a student about three times longer to complete the test than it would take the instructor to do the same test.

The tests are arranged so that all the answers are placed and read on the right-hand side of the page. This format makes it easier for grading the paper because all answers will be at the right-hand side of the page.

A teacher's answer key is included in the Instructor's Guide, which is available from the publishers. In some cases, the wording of answers given by the students will vary from those of the authors; therefore, the teacher should use judgment in deciding whether the student's answer is proper and mark it accordingly.

The tests should be a good source of information to the instructor, showing the areas where students do not understand or are having difficulties. These points should be clarified by the instructor when the test is being reviewed.

While the text has been written to be used as a teaching and learning tool, its success will depend on the individual differences and background experiences of the teachers. There is no one teaching method that is guaranteed to work well for everyone. Teachers should use the method of instruction that is most effective and with which they feel most comfortable. To make lessons effective and as interesting as possible to the students, it is advisable to use as many visuals, models, videotapes, and teaching aids as possible, and actively involve students in the learning process.

# A GUIDE TO THE USE OF THIS WORKBOOK

The questions in this book are for use with the text *Machine Tool and Manufacturing Technology*. Five different types or categories of questions are used: completion, true or false, multiple choice, matching, and identification. Sample questions and answers are provided to assist the reader in answering them properly. Please study these example questions carefully to understand how each type of question should be answered.

## Type 1—Sentence Completion

These questions contain an underlined blank space(s) that is identified with a raised question mark. Record the correct answer(s) in the blank space(s) provided at the right-hand side of the page.

Example:                                                                                    Answer

1. In training programs, it is common practice to machine           1.    lathe

   workpieces between __?__ centers.

The correct answer is **lathe** and it should be written in the space provided at the right-hand column as shown.

## Type 2—True or False

For each statement, determine whether the statement is true or false and circle the correct letter (**T** for true or **F** for false) in the right-hand column.

Example:                                                                                    Answer

2. The Bronze Age followed the Stone Age.                              2.    (T)    F

This statement is true so the letter **T** should be circled.

## Type 3—Multiple Choice

There are four different choices provided for each question. Select the answer you feel is most correct and circle the letter in the right-hand column which indicates your choice.

Example:                                                                                    Answer

3. Which of the following is considered a layout           3.    A    B    (C)    D

   accessory?

   (a) micrometer            (c) parallels

   (b) slide rule            (d) optical scanner

The correct answer is **parallels**, therefore the letter C should be circled in the right-hand column.

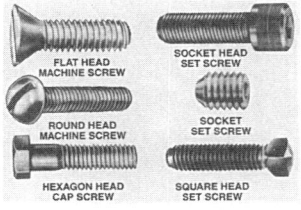

## Type 4—Matching

These texts consist of photographs, drawings or illustrations, each identified by a letter on the left-hand side of the paper. In the right-hand column, information is provided to identify the visual. Match the letter relating to the visual with the proper information.

Example:                                                                 Answer

4.  [file illustration labeled TANG, LENGTH, HEEL, FACE, EDGE, POINT]    4. Used in assembly work       B
    A

5.  [illustration of various screws: FLAT HEAD MACHINE SCREW, SOCKET HEAD SET SCREW, ROUND HEAD MACHINE SCREW, SOCKET SET SCREW, HEXAGON HEAD CAP SCREW, SQUARE HEAD SET SCREW]    5. Used in machine shop work    A
    B

## Type 5—Identification

This type of test contains photographs or labeled visuals, on which parts must be identified by placing the appropriate name or letter in the right-hand column.

Example:

6. Record the parts of the lathe indicated by the letters on the illustration.

                                                    Answers
                                                    6. A  saddle
                                                       B  compound rest
                                                       C  compound rest screw handle
                                                       D  cross-slide

A Guide to the Use of this Workbook

# TEST #1
# TECHNICAL DRAWINGS/PRINTS (SECTION 3, UNIT 3)

Technical drawings and prints are used by industry to describe a part or an assembly in enough detail so that it can be made by a machinist or toolmaker. It is the common language of industry and uses lines, dimensions, symbols, and views to describe a part.

Place the correct word(s) in the space(s) at the right-hand side of the page to make the statement complete and true.

ANSWER

1. An __?__ drawing shows how the individual components of a product are put together.

    1. _____

2. A __?__ drawing shows a single part's size, shape, dimensions, surface finish, etc.

    2. _____

3. There are __?__ main systems for showing dimensions on technical drawings.

    3. _____

4. For computer numerical control work, a part may be dimensioned in the __?__ system or the __?__ system.

    4. _____
    _____

5. Another term for the orthographic view is the __?__ method.

    5. _____

6. Orthographic projection shows three views of an object, the __?__ , __?__ , and __?__ - __?__ view.

    6. _____
    _____
    _____-_____

7. Each view in the orthographic view may be seen by looking __?__ at each of the surfaces.

    7. _____

8. Cylindrical parts are generally shown on prints in two views, the __?__ and __?__ __?__ .

    8. _____
    _____-_____

9. Complicated interior forms may be shown more clearly by a __?__ view.

    9. _____

10. The American National Standards Institute (ANSI) have standardized drafting lines which are known as the __?__ of lines.

    10. _____

11. When using allowances and tolerances, __?__ size is the size from which limits of size are based.

    11. _____

12. __?__ is the intentional difference in the size of mating parts.

    12. _____

13. __?__ fit is the space a part may rotate or move in relation to a moving part.

    13. _____

Test #1 Technical Drawings/Prints (Section 3, Unit 3)

14. When two parts are forced together to act as a single piece, it is called a __?__ fit.

14. _____

15. __?__ represent the largest and smallest sizes of a part that are permissible.

15. _____

16. When a part is too large to draw to actual size, __?__ size is used.

16. _____

_____

Total—20 Marks

# TEST #2
# MACHINING PROCEDURES (SECTION 3, UNIT 4)

Manufacturing processes have been developed which, when followed, can best use human resources, materials and machine tools. This ensures that high quality parts will be made more quickly and accurately at the lowest possible cost.

Consider whether each statement is true or false, then indicate your choice by circling the proper letter in the right hand column.

ANSWER

1. Much of the work produced in the machine shop is round and is machined on a lathe or a turning center.    1. T F

2. In industry, much of the round work is held in a lathe chuck.    2. T F

3. In training programs, most work is held in a lathe chuck.    3. T F

4. It is considered good practice to first rough turn all diameters before finish turning.    4. T F

5. When turning in a lathe, all small diameters should be machined first.    5. T F

6. Finish turning before the workpiece has returned to room temperature will cause an inaccurate measurement.    6. T F

7. Machining between centers allows the workpiece to be removed from the machine and returned to the same accuracy.    7. T F

8. Working between lathe centers will prevent the entire work length to be machined.    8. T F

9. It is important to drill center holes in each end when machining round workpieces held in a lathe chuck.    9. T F

Place the correct word(s) in the space(s) at the right-hand side of the page to make the statement complete and true.

ANSWER

10. Work in a lathe chuck must be held short for the purpose of __?__ and to prevent __?__.    10. _____ _____

11. Never extend work beyond the chuck jaws more than __?__ times its __?__ unless it is supported.    11. _____ _____

12. Flat material should generally be cut approximately __?__ larger than the finish size required.    12. _____

Test #2 Machining Procedures (Section 3, Unit 4)

13. Work surfaces should be coated with __?__ dye in preparation for a layout.

14. The outline of the part should be laid out according to the __?__ on the print.

15. Layout lines can be preserved by indenting them with light __?__ marks.

16. Large sections of the workpiece should be removed quickly on a contour __?__ to within __?__ of the layout lines.

17. A __?__ should be used to scribe the reference circles for all hole locations.

13. _____

14. _____

15. _____

16. _____
    _____

17. _____

_____

Total—20 marks

# TEST #3
# ACCIDENT PREVENTION (SECTION 4, UNIT 5)

Safety is everyone's business and also everyone's responsibility. Accidents don't just happen; they are generally caused by carelessness on someone's part and can be avoided. Learn to work safely and observe all safety regulations.

Place the correct word(s) in the space(s) at the right-hand side of the page to make the statement complete and true.

ANSWER

1. The __?__ way is considered the most correct and efficient way of working.

   1. _____

2. Most accidents are caused by __?__ in work habits, or by __?__ .

   2. _____
   _____

3. Three types of eye protection used in machine shops are: __?__ __?__ , __?__ __?__ , and __?__ __?__ .

   3. ____ ____
   ____ ____
   ____ ____

4. Safe work habits can be developed by being __?__ and __?__ at all times and also being dressed __?__ .

   4. _____
   _____
   _____

5. Never wear __?__ clothing because it can be __?__ in revolving parts of a machine.

   5. _____
   _____

Consider whether each statement is true or false, then indicate your choice by circling the proper letter in the right-hand column.

ANSWER

6. Never wear loose clothing when operating machinery.     6. T   F
7. Never wear a necktie while working in a machine shop.     7. T   F
8. Loosely woven material such as sweaters may be worn in a machine shop.     8. T   F
9. Short-sleeved clothing should never be worn.     9. T   F
10. Wrist watches, rings, bracelets, or other jewelry should never be worn in a machine shop.     10. T   F
11. Gloves may be worn when operating machinery.     11. T   F
12. Canvas shoes offer some protection against sharp chips or falling objects.     12. T   F

| | | | | | |
|---|---|---|---|---|---|
| 13. | One of the most common and painful accidents on a drill press is getting chips caught in a rotating drill. | 13. | T | F |
| 14. | Never attempt to clean a machine until it has come to a complete stop. | 14 | T | F |
| 15. | A cloth may be used to remove chips from a machine. | 15. | T | F |
| 16. | Never use compressed air to blow away chips from a machine. | 16. | T | F |
| 17. | Always keep hands away from moving parts. | 17. | T | F |
| 18. | Only one person should operate a machine at any one time. | 18. | T | F |
| 19. | Safe lifting practices permit only one person to lift heavy objects. | 19. | T | F |

Total—25 marks

# TEST #4
# MEASUREMENT SYSTEMS (SECTION 5, UNIT 6)

The world has depended on some form of measurement system since the beginning of civilization. In ancient times, parts of the human body were used, and through the ages, new measuring tools continued to make more accurate measurements possible. Today, tools capable of measuring in millionths of an inch are commonplace.

Place the correct word(s) in the space(s) at the righthand side of the page to make the statement complete and true.

ANSWER

1. __?__ is another name for the science of precision measurement.

    1. _____

2. __?__ measuring tools, __?__-__?__ gaging, __?__ measuring systems and __?__ non-contact tools are capable of measuring in millionths of an inch.

    2. _____
    _____-_____
    _____
    _____

3. The __?__ and __?__ systems are the two major measurement systems in the world.

    3. _____
    _____

4. Over __?__ % of the countries in the world are presently using the __?__ system.

    4. _____
    _____

5. As of 1983, the meter is defined as the distance traveled by __?__ in a __?__ during 1/299,792,458 of a second.

    5. _____
    _____

6. All multiples and subdivision of the meter are directly related to the meter by a factor of __?__ .

    6. _____

7. The inch steel rule is generally divided into __?__ graduations.

    7. _____

8. Metric rules are usually graduated in __?__ and __?__-__?__ .

    8. _____
    _____-_____

9. On inch steel rules, dimensions of __?__ are about as small as can be seen without the aid of a magnifying glass.

    9. _____

10. When beginning to measure with a steel rule, it is advised to place the __?__ or __?__ graduation line on the edge of the work.

    10. _____
    _____

Test #4 Measurement Systems (Section 5, Unit 6)

11. __?__-__?__ rules are used for measuring narrow spaces or hard-to-reach surfaces.

11. _____-_____

12. __?__ rules are often used to make linear measurements less than 1/64 in.

12. _____

_____

Total—20 marks

# TEST #5
# MICROMETERS (SECTION 5, UNIT 7)

The outside micrometer is the most widely-used measuring instrument in metalworking shops. Various micrometers are available and they are capable of measuring within .0001 in. in the inch system and 0.002 mm in the metric system.

Consider whether each statement is true or false, then indicate your choice by circling the proper letter in the right-hand column.

|    |                                                                                                              |     | ANSWER |   |
|----|--------------------------------------------------------------------------------------------------------------|-----|--------|---|
| 1. | Both the inch micrometer and the metric micrometer are similar in construction.                              | 1.  | T      | F |
| 2. | The pitch of the spindle screw is the only difference between these micrometers.                             | 2.  | T      | F |
| 3. | The anvil is the movable measuring face.                                                                     | 3.  | T      | F |
| 4. | Measurements are made between the faces of the thimble and the spindle on a micrometer.                      | 4.  | T      | F |
| 5. | On standard inch micrometers, measurements as fine as .001 in. are possible.                                 | 5.  | T      | F |
| 6. | An inch micrometer can read in ten-thousandths of an inch if it is equipped with a vernier scale.            | 6.  | T      | F |
| 7. | Each division of the vernier scale on the sleeve of a metric micrometer has a value of 0.002 mm.             | 7.  | T      | F |
| 8. | The pitch of a standard one inch micrometer thread is equal to .025 in.                                      | 8.  | T      | F |
| 9. | The pitch of a standard metric micrometer thread is 0.5 mm.                                                  | 9.  | T      | F |
| 10.| There is no difference between the term "center line" and the "index line" that is found on the sleeve.      | 10. | T      | F |

Record the micrometer reading for each illustration in the proper space in the right-hand column. State whether the reading is in in. or mm.

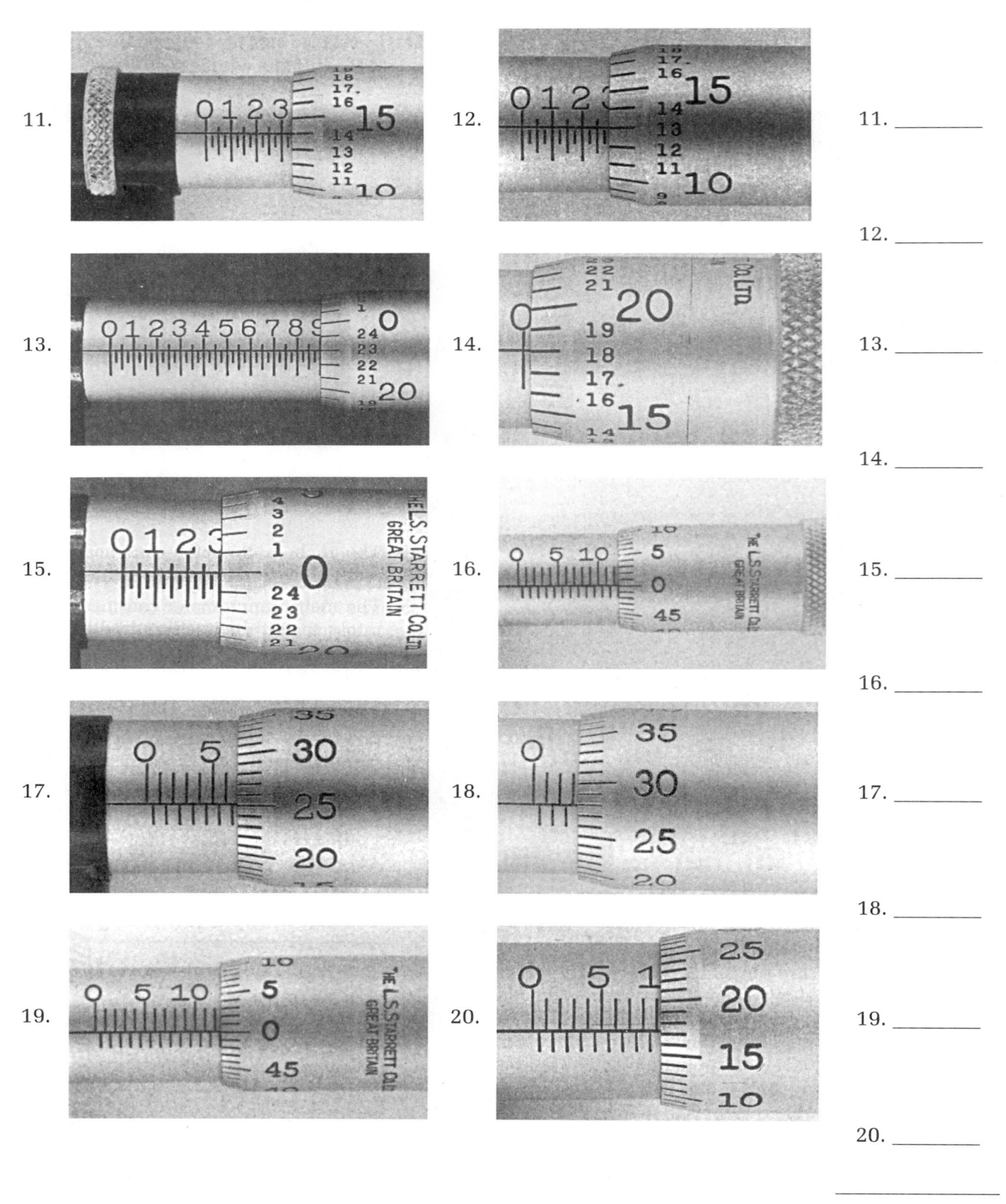

11.

12.

13.

14.

15.

16.

17.

18.

19.

20.

11. _____

12. _____

13. _____

14. _____

15. _____

16. _____

17. _____

18. _____

19. _____

20. _____

Total—20 marks

14   Lab Manual to Accompany Machine Tool and Manufacturing Technology

# TEST #6
# VERNIER CALIPERS (SECTION 5, UNIT 8)

The vernier caliper is a precision measuring instrument used to make inside and outside measurements. They are available for both inch and metric measurements, and are generally used where a micrometer would not be suitable.

Place the name of each part of this 50-division vernier caliper in the proper space in the right-hand column.

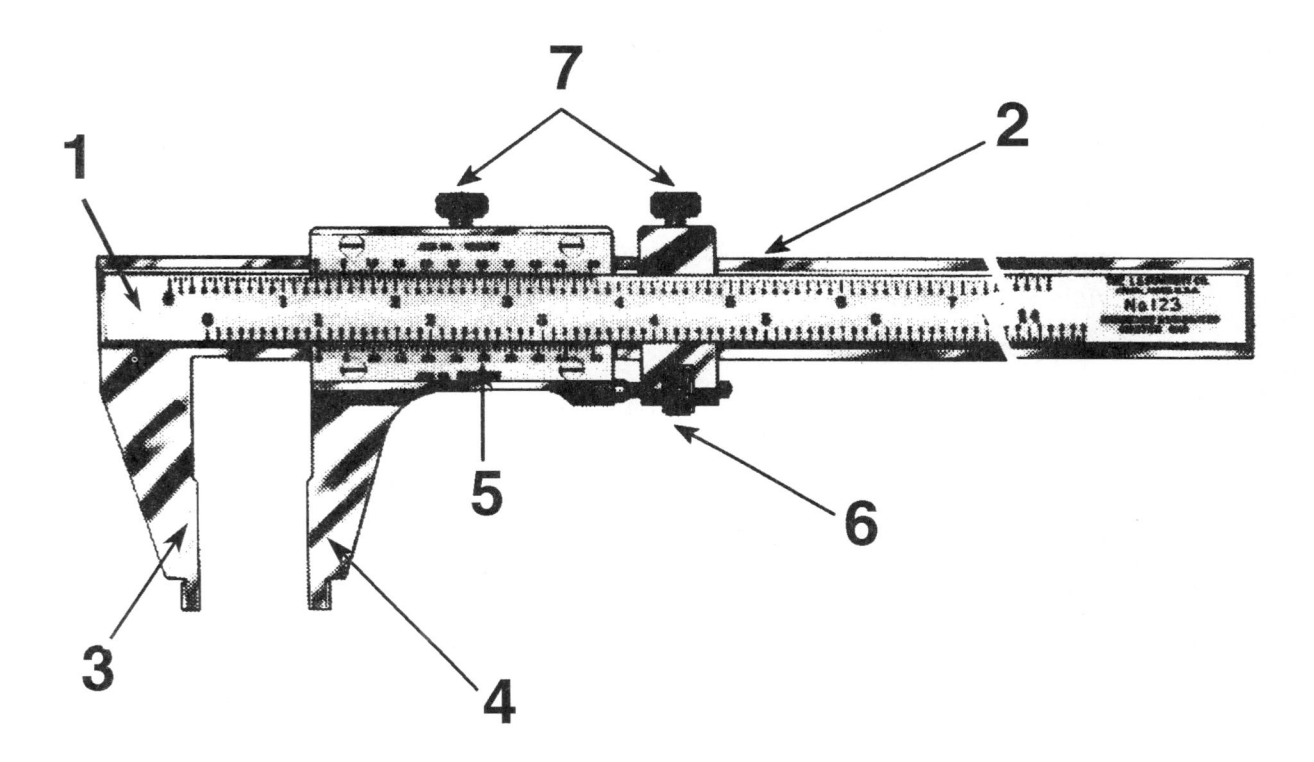

ANSWER

1. _____

2. _____

3. _____

4. _____

5. _____

6. _____

7. _____

Test #6 Vernier Calipers (Section 5, Unit 8)                                                                                15

Select the correct answer for each question and circle the letter in the right-hand column indicating your choice.

ANSWER

8. Each graduation on the sliding-jaw scale on an inch vernier caliper has a value of

    (A) .0001 in.
    (B) .001 in.
    (C) .0015 in.
    (D) .002 in.

    8.  A   B   C   D

9. The small numbers on the bar of a 50-division inch vernier caliper represents

    (A) 1.000 in.
    (B) .100 in.
    (C) .010 in.
    (D) .001 in.

    9.  A   B   C   D

10. Each division on the bar of a 50-division inch vernier caliper has a value of

    (A) .001 in.
    (B) .010 in.
    (C) .050 in.
    (D) .500 in.

    10. A   B   C   D

11. The indicator face of a direct reading inch dial caliper is usually divided into

    (A) 10 graduations
    (B) 50 graduations
    (C) 100 graduations
    (D) 200 graduations

    11. A   B   C   D

12. The value of each indicator inch dial graduation is

    (A) .0001 in.
    (B) .0005 in.
    (C) .001 in.
    (D) .0015 in.

    12. A   B   C   D

13. Each graduation on the vernier scale of a 50-division metric vernier caliper has a value of

    (A) 0.001 mm
    (B) 0.01 mm
    (C) 0.02 mm
    (D) 0.03 mm

    13. A   B   C   D

14. Each numbered division on a 50-division metric vernier caliper has a value of

    (A) 1 mm  (C) 50 mm
    (B) 10 mm  (D) 100 mm

    14.  A  B  C  D

15. Each division on the bar of a 50-division metric vernier caliper has a value of

    (A) 0.01 mm  (C) 0.05 mm
    (B) 0.02 mm  (D) 1 mm

    15.  A  B  C  D

16. The indicator face of a metric dial caliper is usually divided into

    (A) 10 graduations  (C) 100 graduations
    (B) 50 graduations  (D) 200 graduations

    16.  A  B  C  D

17. The value of each indicator metric dial graduation is

    (A) 0.05 mm  (C) 0.02 mm
    (B) 0.01 mm  (D) 0.03 mm

    17.  A  B  C  D

Record the vernier caliper reading for each illustration in the proper space in the right-hand column.

18. 

18. _____

19. 

19. _____

20. 

20. _____

Total—20 marks

Test #6 Vernier Calipers (Section 5, Unit 8)

# TEST #7
# INSPECTION TOOLS (SECTION 5, UNIT 9)

Accurate measurement and inspection systems make it possible to produce high quality goods which are acceptable throughout the world. One of the more recent inspection systems is the laser which makes it possible to check the accuracy of a part while it is being machined, thereby almost eliminating scrap parts.

Place the correct word(s) in the space(s) at the right-hand side of the page to make the statement complete and true.

ANSWER

1. Gage blocks are physical standards to which all measuring tools are __?__ or __?__ .

    1. _____
       _____

2. Three reasons why gage blocks are used in machine shop work are __?__ , __?__ , and __?__ .

    2. _____
       _____
       _____

3. Alloy steel gage blocks are stabilized through alternate cycles of extreme __?__ and __?__ .

    3. _____
       _____

4. The Class AA gage block set is commonly called a laboratory or __?__ set.

    4. _____

5. The Class A set is used for __?__ purposes.

    5. _____

6. The Class B set is commonly called the __?__ set.

    6. _____

7. The operation of __?__ ensures that quality is manufactured into a product.

    7. _____

8. __?__ and __?__ - __?__ measuring tools are two types of electro-optical inspection measurement tools.

    8. _____
       _____-_____

9. __?__ measuring tools are those that come into physical contact with the part during the measuring or inspection process.

    9. _____

10. In-process gaging provides __?__ feedback to the control of the machine to stop the operation when the part is to __?__ .

    10. _____
        _____

11. The __?__ is an electronic measuring machine used for advanced, multi-purpose quality control.

    11. _____

12. __?__ systems are designed for 100% inspection of the part or component during the production operation.

    12. _____

13. __?__-__?__ measuring tools do not come into physical contact with the part in the manufacturing or inspection process.

13. _____-_____

14. Optical gaging systems can accurately measure parts to millionths of an inch while the part is being electronically __?__.

14. _____

_____
Total—20 marks

# TEST #8
# LAYING OUT (SECTION 6, UNIT 10)

Laying out lines or circles on the surface of a workpiece is often done to guide the machinist to indicate the finished size and shape of a part. The care and accuracy of the layout plays an important role in determining the accuracy of the finished workpiece.

Select the correct answer for each question and indicate your choice by circling the correct letter in the right-hand column.

ANSWER

1. A machinist should be able to

   (A) read prints

   (B) select and use layout tools

   (C) transfer dimensions to a part

   (D) perform all these tasks

   1. A B C D

2. A semi-precision layout is one where the accuracy of the work is

   (A) more than 1/64 in. (0.38 mm)

   (B) less than 1/64 in. (0.38 mm)

   (C) more than 1/32 in. (0.96 mm)

   (D) none of these

   2. A B C D

3. A precision layout is one where the accuracy of the work is

   (A) more than 1/64 in. (0.38 mm)

   (B) less than 1/64 in. (0.38 mm)

   (C) more than 1/32 in. (0.96 mm)

   (D) none of these

   3. A B C D

4. The most common layout coating used in the machine shop is

   (A) layout dye         (C) chalk

   (B) blue vitriol       (D) heat treated bluing

   4. A B C D

5. Surface plates may be made of

   (A) cast iron          (C) ceramic

   (B) granite            (D) all of these

   5. A B C D

6. Layout lines on metal surfaces are drawn by                     6.   A   B   C   D

   (A) fine chisel              (C) scriber
   (B) square                   (D) prick punch

7. A divider is used for                                           7.   A   B   C   D

   (A) transferring measurement
   (B) comparing distances
   (C) scribing arcs and circles
   (D) all of these

8. The purpose of a prick punch is to                              8.   A   B   C   D

   (A) scribe lines
   (B) draw lines
   (C) produce witness marks
   (D) make holes

9. The included angle of a prick punch point is                    9.   A   B   C   D

   (A) 45 - 60°                 (C) 60 - 90°
   (B) 30 - 60°                 (D) 90 - 110°

10. The included angle of a center punch point is                  10.  A   B   C   D

    (A) 30°                     (C) 60°
    (B) 45°                     (D) 90°

Consider whether each statement is true or false, then indicate your choice by circling the proper letter in the right-hand column.

|     |                                                                                      |     | ANSWER |
| --- | ------------------------------------------------------------------------------------ | --- | ------ |
| 11. | A combination set contains a bevel protractor, steel rule, square head, and a center head. | 11. | T   F |
| 12. | The center head has an included angle of 60°.                                         | 12. | T   F |
| 13. | The center head may be used for locating the centers on the ends of octagonal stock.  | 13. | T   F |
| 14. | A bevel protractor on a combination set has a range of 180°                           | 14. | T   F |
| 15. | A surface gage may be used for a precision layout.                                    | 15. | T   F |

16. A vernier height gage is considered a precision instrument.     16. T    F

17. A vernier height gage can measure to within .001 in. (0.02 mm) when equipped with a dial indicator.     17. T    F

18. An angle plate has all its surfaces ground square and parallel to a 90° angle.     18. T    F

19. Parallels do not come in matched pairs.     19. T    F

20. A template is a master pattern that can be used as a guide to lay out a number of workpieces.     20. T    F

Total—20 marks

# REVIEW TEST #1
# MEASUREMENT AND LAYOUT
# (SECTIONS 5 AND 6—UNITS 6 TO 10)

**PART 1**

Place the letter of each measuring tool or system in the proper space in the right-hand column.

| | ANSWER |
|---|---|
| 1. | combination micrometer _____ |
| 2. | gage blocks _____ |
| 3. | electronic caliper _____ |
| 4. | combination set _____ |
| 5. | surface plate _____ |
| 6. | vernier caliper _____ |
| 7. | CMM _____ |
| 8. | metric micrometer _____ |
| 9. | decimal rule _____ |
| 10. | angle plate _____ |
| 11. | hook rule _____ |
| 12. | inch dial caliper _____ |
| 13. | dial/digital height gage _____ |
| 14. | short length rule _____ |

Review Test #1 Measurement and Layout (Sections 5 and 6—Units 6 to 10)    25

O

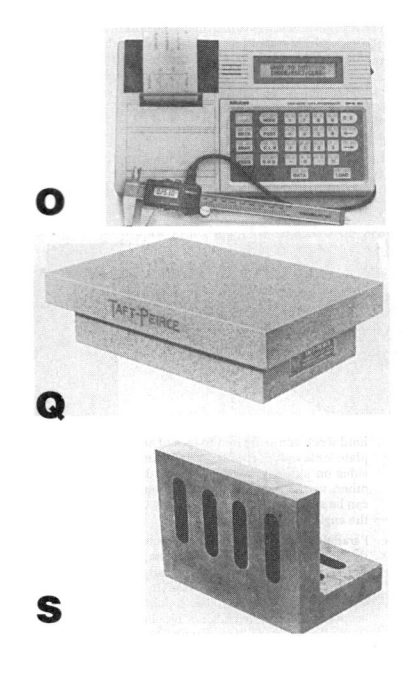

P

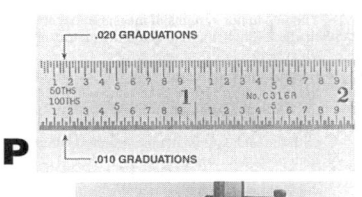

Q

R

S

T

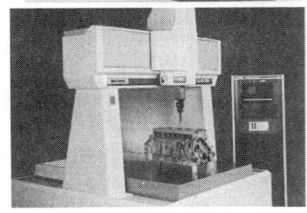

15. standard inch rule  _____

16. metric rule  _____

17. metric dial caliper  _____

18. vernier gage blocks  _____

19. adjustable square  _____

20. inch micrometer  _____

**PART 2**

Place the measuring tool reading for each illustration in the proper space in the right-hand column.

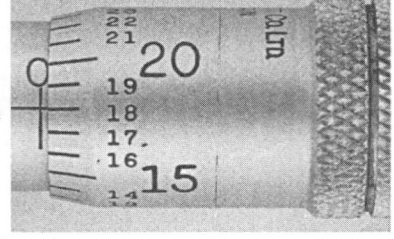

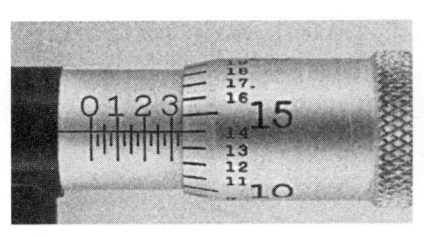

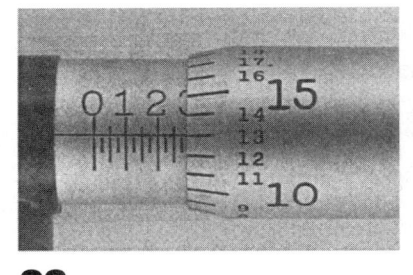

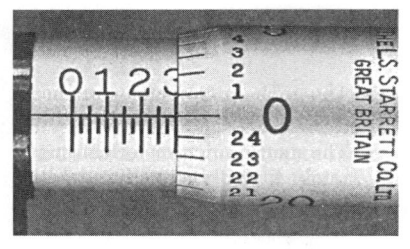

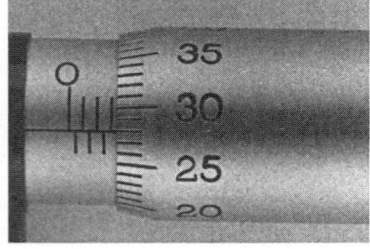

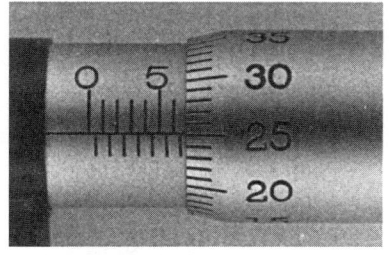

ANSWER

21

22

23

24

25

26

21. _____

22. _____

23. _____

24. _____

25. _____

26. _____

26  Lab Manual to Accompany Machine Tool and Manufacturing Technology

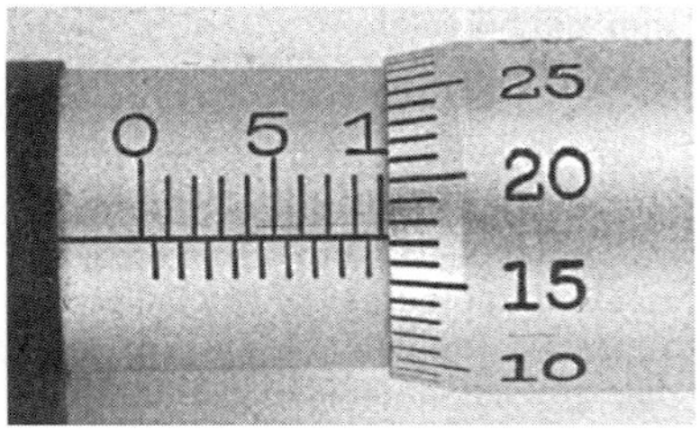

**27**

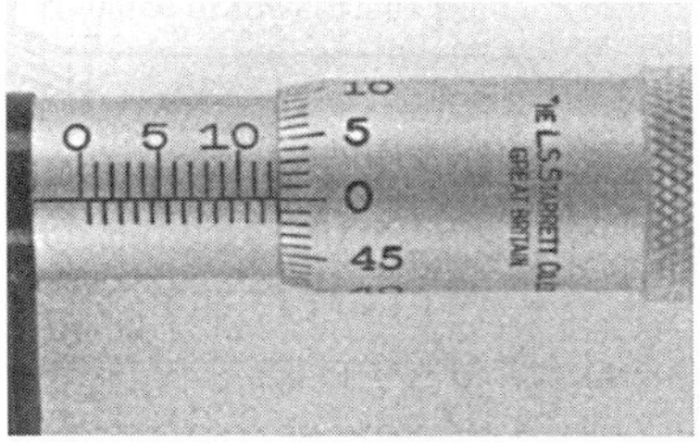

**28**

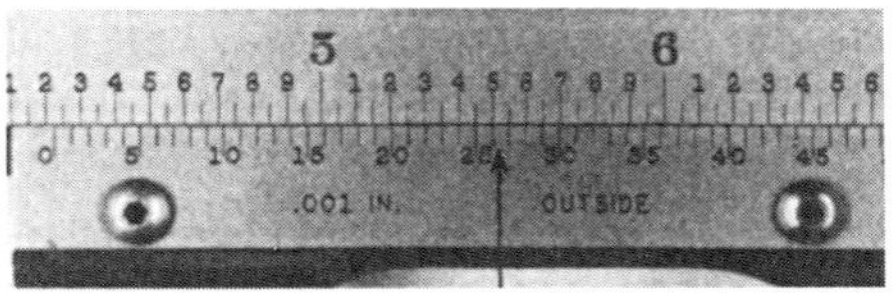

**29**

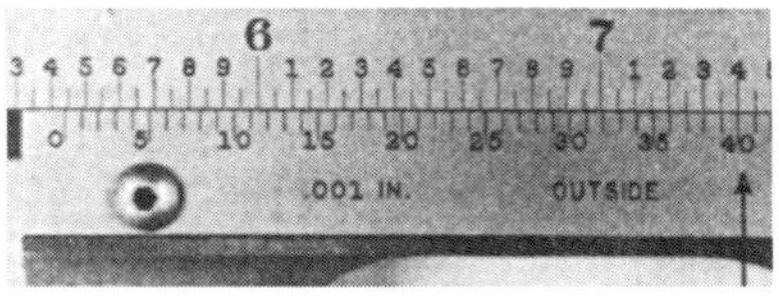

**30**

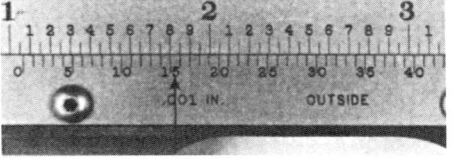

**31**

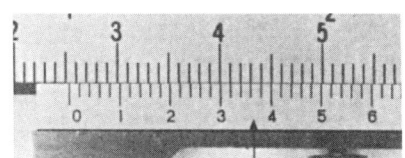

**32**

27. _____

28. _____

29. _____

30. _____

31. _____

32. _____

## PART 3

Place the correct word(s) in the space(s) at the right-hand side of the page to make the statement complete and true.

ANSWER

33. The science of precision measurement is also known as __?__ .   33. _____

34. The meter is defined as the __?__ that __?__ travels in a vacuum in a __?__ .   34. _____
_____
_____

35. __?__ __?__ are the acceptable __?__ standards of measurement.   35. _____ _____
_____

36. __?__ __?__ control is the inspection system used in industry to prevent errors and improve product quality.   36. _____
_____

37. Precision contact measuring tools used for inspection purposes are generally __?__ or a combination of __?__-__?__ .   37. _____
_____ _____

38. Contact measuring tools should only be __?__ when the machining operation is __?__ .   38. _____
_____

Review Test #1 Measurement and Layout (Sections 5 and 6—Units 6 to 10)

39. Scanning __?__ systems use a thin beam of __?__ which scans the part or measurement area at a constant speed.

40. Television cameras are being used for some forms of __?__-__?__ measurement.

41. The ultimate goal of improved product quality must be that the __?__ is completely __?__.

39. _____
    _____

40. _____-_____

41. _____
    _____

Total—50 marks

# TEST #9
# MANUFACTURE OF IRON AND STEEL (SECTION 7, UNIT 11)

Of all the metals produced, none are more important than iron and steel. Iron and steel products are found in, or are used to produce, almost every product used by humans. A good understanding of various metals and their properties is important to anyone in metalworking.

Consider whether each statement is true or false, then indicate your choice by circling the proper letter in the right-hand column.

ANSWER

1. Iron consists of approximately 5% of the earth's crust. — 1. T F
2. Steel is the most versatile of metals. — 2. T F
3. Iron ore is the chief raw material used in the manufacture of iron and steel. — 3. T F
4. Iron ore, coal, and limestone are used to produce pig iron. — 4. T F
5. Most iron ore is mined by open pit mining. — 5. T F
6. Underground or shaft mining is more economical than open pit mining. — 6. T F
7. Hematite is a rich ore containing about 50% iron. — 7. T F
8. Pelletizing increases the iron concentration of low grade iron ore. — 8. T F
9. Coke is made from a special grade of soft coal. — 9. T F
10. Limestone is a by-product of coal. — 10. T F
11. The direct ironmaking process produces iron in a one-step process. — 11. T F
12. Pig iron is used to manufacture cast iron. — 12. T F
13. The Bessemer furnace is still used in the steelmaking process. — 13. T F
14. The electric arc furnace is used to produce low carbon steels. — 14. T F
15. The direct steelmaking process involves the operation of smelting, pre-reduction, offgas cleaning, and refining. — 15. T F

Place the correct word(s) in the space(s) at the right-hand side of the page to make the statement complete and true.

ANSWER

16. Hot ingots of steel are placed in __?__ pits to bring them to uniform temperatures before being sent to rolling mills. — 16. _____

17. Continuous or __?__ casting is an efficient method of converting molten steel into semi-finished shapes.

17. _____

18. Semi-finished shapes produced by continuous casting include __?__ , __?__ , and __?__ .

18. _____
_____
_____

19. Minimills are __?__ expensive and more __?__ than the electric furnace steel-making process.

19. _____
_____

20. Minimills are __?__ , __?__ , and more __?__ than larger size integrated mills.

20. _____
_____
_____

_____

Total—25 marks

# TEST #10
# METALS AND THEIR PROPERTIES (SECTION 7, UNIT 12)

The composition and properties of ferrous metals may be changed by adding various alloying elements during manufacture. These alloying elements can increase tensile strength, reduce weight, increase corrosion resistance, and give steels better hardening qualities.

Place the correct word(s) in the space(s) at the right-hand side of the page to make the statement complete and true.

ANSWER

1. The property of a metal which permits no permanent distortion before breaking is called __?__ .  1. _____

2. __?__ is the ability of the metal to be permanently deformed without breaking.  2. _____

3. Elasticity is the ability of the metal to __?__ to its original shape after an outside force has been __?__ .  3. _____  _____

4. Hardness may be defined as the __?__ to penetration or plastic __?__ .  4. _____  _____

5. __?__ is the property of a metal which permits it to be hammered or rolled into other sizes and shapes.  5. _____

6. Toughness is the property of a metal to withstand __?__ or __?__ .  6. _____  _____

7. Fatigue-failure is the point at which metal __?__ , __?__ , or __?__ as a result of repeated stress.  7. _____  _____  _____

8. The three general classes of ferrous metals are steel, __?__ iron and __?__ iron.  8. _____  _____

9. Steels are generally classified by their __?__ content.  9. _____

10. Alloy steels contain two or more metals to give steel new __?__ .  10. _____

11. Aluminum is made from __?__ ore.  11. _____

12. Nonferrous metals are metals which contain little or no __?__ .  12. _____

Test #10 Metals and Their Properties (Section 7, Unit 12)

13. Brass and bronze are examples of __?__ alloys.        13. _____

14. __?__ metals may be identified by spark testing.      14. _____

                                                          _____
                                                          Total—20 marks

# TEST #11
# METAL CUTTING (SECTION 8, UNIT 13)

The continual development of new metals and alloys makes it important to know what cutting tools and conditions are required to machine them properly. This knowledge will prolong the life of cutting tools and produce accurate workpieces.

Place the correct word(s) in the space(s) at the right-hand side of the page to make the statement complete and true.

ANSWER

1. The original theory of what happens when a tool cuts metal was thought to be similar to what occurs when an __?__ splits __?__ .

    1. _____
       _____

2. Through research, it was found that when a tool enters metal, it causes metal in front of it to become __?__ , __?__ , and deform.

    2. _____
       _____

3. Cutting fluid is applied to the chip-tool interface to reduce __?__ and __?__ the tool and the workpiece.

    3. _____
       _____

4. A layer of compressed metal which sticks to and piles up on a cutting tool is called a __?__-__?__ edge.

    4. _____-_____

5. Three types of chips produced during machining are __?__ , __?__ , and continuous with a built-up edge.

    5. _____
       _____

6. Machinability is a term that describes the __?__ or __?__ with which metal can be machined.

    6. _____
       _____

7. __?__-__?__ steel tools are commonly used in school shops because of their general-purpose use and relatively low cost.

    7. _____-_____

8. Most ceramic cutting tools are manufactured from __?__ oxide.

    8. _____

9. Manufactured diamond and cubic boron nitride (CBN) cutting tools are classed as __?__ .

    9. _____

10. Cubic boron nitride (CBN) is used to machine and grind hard, abrasive __?__ materials.

    10. _____

11. Manufactured diamond is used to machine and grind hard, abrasive __?__ materials.

    11. _____

12. The main properties of manufactured diamond and CBN are __?__ , __?__ __?__ , __?__ __?__ , and __?__ __?__ .

12. _____
   _____
   _____
   _____
   _____
   _____
   _____

13. Polycrystalline cutting tools are manufactured by a __?__ - __?__ , __?__ - __?__ process which fuses diamond or cubic boron nitride to a cemented tungsten carbide base.

13. _____-_____
    _____-_____

14. The life of any cutting tool generally depends on the __?__ and __?__ which occurs at the cutting edge.

14. _____
    _____

15. One of the characteristics of super-abrasive tools is that the machining heat is quickly __?__ from the chip-tool interface.

15. _____

_____

Total—25 marks

# TEST #12
# CUTTING FLUIDS (SECTION 8, UNIT 14 )

Cutting fluids are very important to the metalworking industry because they reduce the effects of friction and heat produced while machining. This affects the performance of the cutting tools, increases material-removal rates, improves quality, and increases productivity.

Select the correct answer for each question and indicate your choice by circling the correct letter in the right-hand column.

ANSWER

1. How much of the heat is generated by the external friction of the chip sliding over the cutting tool face?   1.  A  B  C  D

   (A) one-quarter  (C) two-thirds
   (B) one-third    (D) one-half

2. How much of the heat is caused by the resistance of the metal atoms of the workpiece being removed?   2.  A  B  C  D

   (A) one-quarter  (C) two thirds
   (B) one-third    (D) one-half

3. Cutting fluids remove heat from the   3.  A  B  C  D

   (A) cutting tool  (C) workpiece
   (B) chip          (D) all of these

4. In addition to rust and corrosion prevention, a good cutting fluid should also have   4.  A  B  C  D

   (A) bacteria control  (C) stability
   (B) nonflammability   (D) all of these

5. The selection and proper application of a cutting fluid should result in   5.  A  B  C  D

   (A) increased labor costs     (C) increased power costs
   (B) increased productivity    (D) none of these

6. Which of the following has a major effect on the coolant quality of water-base cutting fluids?   6.  A  B  C  D

   (A) dye      (C) hardness
   (B) alcohol  (D) oil

Test #12 Cutting Fluids (Section 8, Unit 14 )

7. For best coolant life, the water used in mixing fluid concentrates should be

    (A) filtered      (C) boiled

    (B) deionized      (D) polished

7.   A   B   C   D

8. Water-soluble fluids provide

    (A) corrosion resistance      (C) cooling

    (B) lubrication      (D) all of these

8.   A   B   C   D

Consider whether each statement is true or false, then indicate your choice by circling the proper letter in the right-hand column.

ANSWER

9. Water provides the best cooling because it can absorb and carry away more heat than any other fluid.     9.   T   F

10. Oil is an excellent lubricant, an excellent coolant, but is flammable.     10.   T   F

11. Water soluble fluids cool, lubricate, prevent corrosion, and are non-flammable.     11.   T   F

12. Chemical cutting fluids are stable but do not mix easily with water.     12.   T   F

13. Chemical cutting fluids are generally used for low stock-removal grinding.     13.   T   F

14. Chemical cutting fluids are available in five types.     14.   T   F

15. Emulsions is another name for soluble oils.     15.   T   F

16. There are three types of soluble oils.     16.   T   F

17. Straight cutting oils are classified under two types: active and inactive.     17.   T   F

18. Straight mineral oils provide excellent lubrication and are effective in dissipating heat.     18.   T   F

19. Lard and sperm oils have limited uses as cutting fluids.     19.   T   F

20. Fatty mineral oils provide better surface finishes on ferrous and nonferrous metals.     20.   T   F

21. Sulfochlorinated mineral oils prevent grinding wheel loading and prolong wheel life.   21. T   F

22. The life of the cutting tool or grinding wheel and the efficiency of the metal-removal operation are affected by how the cutting fluid is applied.   22. T   F

23. The most effective way of applying fluids in drill press operations is through the use of oil-feed drills.   23. T   F

24. In any type of milling operation, only one coolant nozzle should be used.   24. T   F

25. In cylindrical grinding, the entire wheel/work contact area should be flooded with a steady stream of cutting fluid.   25. T   F

Total—25 marks

# REVIEW TEST #2
# METALS AND PRINCIPLES OF METAL CUTTING
# (SECTIONS 7 AND 8—UNITS 11 TO 14)

## PART 1

Consider whether each statement is true or false, then indicate your choice by circling the proper letter in the right-hand column.

|     |     | ANSWER |
| --- | --- | --- |
| 1. Steel is one of the most abundant elements found on the earth. | 1. | T  F |
| 2. Steel is the most versatile of metals. | 2. | T  F |
| 3. Less than 50% of the iron ore mined is removed by open pit mining methods. | 3. | T  F |
| 4. Coke is made from a special grade of soft coal containing small amounts of phosphorus or sulfur. | 4. | T  F |
| 5. Slag is a by-product that is formed by melted limestone which combines with the impurities from iron ore and coke. | 5. | T  F |
| 6. The aim of the direct ironmaking process is to produce iron in a one-step process. | 6. | T  F |
| 7. Most of the pig iron manufactured in a blast furnace is used to make cast iron products. | 7. | T  F |
| 8. The Bessemer converter is rarely used in the manufacture of steel. | 8. | T  F |
| 9. The direct steelmaking process is considered to be almost pollution-free. | 9. | T  F |
| 10. Continuous casting is an efficient method of converting molten steel into semi-finished shapes. | 10. | T  F |
| 11. Brittleness may be defined as the resistance to forcible penetration. | 11. | T  F |
| 12. Nonferrous metals generally contain iron. | 12. | T  F |
| 13. Aluminum is the most abundant nonferrous metal mined. | 13. | T  F |
| 14. Brass is an alloy composed mainly of copper with large amounts of tin and zinc. | 14. | T  F |
| 15. Ferrous metals can be identified by spark testing. | 15. | T  F |

## PART 2

Place the correct word(s) in the space(s) at the right-hand side of the page to make the statement complete and true.

ANSWER

16. It was formerly thought that the __?__ of wood by an __?__ accurately defined the cutting __?__ of metal being cut.

16. _____
    _____
    _____

17. In order for cutting tools to remove metal efficiently, they must be __?__ , __?__-__?__ , withstand __?__ , penetrate the work __?__ and maintain their cutting edge during the machining operation.

17. _____
    _____-_____
    _____
    _____

18. In most machining operations, cutting fluid is applied to the chip-tooth __?__ to reduce __?__ and cool the __?__ and the workpiece.

18. _____
    _____
    _____

19. The __?__ chip produces a ribbon-like __?__ of metal and is considered __?__ for efficient cutting action.

19. _____
    _____
    _____

20. __?__ deformation and __?__ elongation is the distortion which occurs to the structure of the metal during a machining operation.

20. _____
    _____

21. The cutting tool __?__ angle affects the __?__ angle, the plane in which the __?__ material separates from the work material.

21. _____
    _____
    _____

22. Two superabrasives are __?__ and __?__ .

22. _____
    _____

23. Polycrystalline diamond (PCD) is used for cutting __?__ , __?__ , __?__ materials.

23. _____
    _____
    _____

24. Polycrystalline cubic boron nitride (PCBN) is used for cutting hard, __?__ , __?__ materials.

24. _____
    _____

# PART 3

Select the correct answer for each question and indicate your choice by circling the correct letter in the right-hand column.

25. When selecting a cutting fluid, which of the characteristics is considered the most important?
    (A) cutting
    (B) noncutting
    (C) both A and B
    (D) neither A nor B
    
    25.  A  B  C  D

26. The most important function of a good cutting fluid is
    (A) cooling
    (B) lubrication
    (C) coating
    (D) both A and B
    
    26.  A  B  C  D

27. Which of the following cutting fluids is the best for cooling?
    (A) oil
    (B) water
    (C) chemical
    (D) emulsions
    
    27.  A  B  C  D

28. How wide should the opening of the fluid supply nozzle be in relation to the cutting tool face for it to be effective?
    (A) three-quarters
    (B) one-quarter
    (C) two-thirds
    (D) three-fifths
    
    28.  A  B  C  D

29. How many major types of cutting fluids are there?
    (A) four
    (B) five
    (C) three
    (D) six
    
    29.  A  B  C  D

30. Which of the following is the best preventive for bacterial growth in cutting fluids?
    (A) low viscosity
    (B) corrosion additive
    (C) synthetic additive
    (D) good housekeeping
    
    30.  A  B  C  D

31. In which of the following machine tool operations can oil feed drills be effectively used?
    (A) drill press
    (B) lathe
    (C) milling machine
    (D) all of these
    
    31.  A  B  C  D

Review Test #2 Metals and Principles of Metal Cutting (Sections 7 and 8—Units 11 to 14)

32. Which of the following cutting fluid application methods is recommended for cylindrical grinding operations?

   (A) through-the-wheel  (C) nozzle flooding
   (B) mist spray  (D) none of these

   32.  A   B   C   D

33. Which of the following cutting fluid applications is used for surface grinding operations?

   (A) flood method  (C) through-the-wheel
   (B) mist spray  (D) all of these

   33.  A   B   C   D

34. Which of the following application methods is recommended for milling machine operations?

   (A) one nozzle  (C) three nozzles
   (B) two nozzles  (D) four nozzles

   34.  A   B   C   D

Total—50 marks

# TEST #13
# HAND TOOLS (SECTION 9, UNIT 15)

Although the use of hand tools in machine shop work has declined because of the electronic and CNC machine tools, they still find limited use for operations such as sawing, filing, polishing, tapping, and threading. Skill in the use of hand tools can only be acquired through years of practice.

Place the correct word(s) in the space(s) at the right-hand side of the page to make the statement complete and true.

ANSWER

1. To keep work from being damaged when it is held in a vise, special vise jaw __?__ are available.  
   1. _____

2. Some jaws are equipped with a __?__ base which allows the vise to be __?__ to any position.  
   2. _____
   _____

3. The ball-peen hammer, or __?__ hammer, is the hammer generally used in machine shop work.  
   3. _____

4. __?__-__?__ hammers are used in assembly and setup to prevent marring the work surface.  
   4. _____-_____

5. A hammer with a __?__ head is dangerous.  
   5. _____

6. __?__ the handle of any hammer if it is cracked.  
   6. _____

7. Never use a hammer when your hands are __?__.  
   7. _____

8. Never strike the face of a hammer against __?__ steel.  
   8. _____

9. The frames on most hacksaws are either flat or tubular, and are __?__ to accommodate different blade lengths.  
   9. _____

10. The distance between each tooth on a hacksaw blade is called the __?__.  
    10. _____

11. An __?__-tooth blade is recommended for general hacksaw use.  
    11. _____

12. In order to provide chip clearance and cut through work quickly, the hacksaw blade should have at least __?__ teeth in contact with the work.  
    12. _____

13. When hacksawing, a speed of about __?__ strokes per minute is recommended.  
    13. _____

14. Files are divided into two classes, __?__-cut, and __?__-cut files.  
    14. _____
    _____

Test #13 Hand Tools (Section 9, Unit 15)  43

15. A file only cuts on the __?__ stroke.            15. _____

16. A file __?__ is used to clean a file and keep it free of chips.            16. _____

17. Never use a file as a __?__ or a hammer.            17. _____

18. To avoid injury to the hand, never file without a __?__ .            18. _____

_____

Total—20 marks

# TEST #14
# THREADING AND REAMING (SECTION 9, UNIT 16)

Both external and internal threads are widely used in the metalworking industry. Threads are used for fastening and assembling parts, transmitting motion, and for measurement purposes.

Place the correct word(s) in the space(s) at the right-hand side of the page to make the statement complete and true.

ANSWER

1. Taps are cutting tools used to cut __?__ threads.　　　　　　1. _____

2. The __?__ of a tap form the cutting edges, provide __?__ for chips, and allow __?__ fluid to enter.　　　　　　2. _____
_____
_____

3. A __?__ tap, tapered approximately six threads from the end, is used to start a thread easily.　　　　　　3. _____

4. A __?__ tap, chamfered at the end for one thread, is used to complete threading a blind hole.　　　　　　4. _____

5. A __?__ tap, tapered approximately three threads from the end, is used for general-purpose tapping.　　　　　　5. _____

6. A tap drill leaves enough material in the hole for the tap to produce __?__ percent of a full thread.　　　　　　6. _____

7. On inch system taps, the __?__ diameter, __?__ of threads per inch, and __?__ of thread are usually found stamped on the shank of the tap.　　　　　　7. _____
_____
_____

8. Metric taps are identified with the letter __?__ followed by the __?__ diameter of the thread in millimeters times the __?__ in millimeters.　　　　　　8. _____
_____
_____

9. Threading dies are used to cut __?__ threads on round work.　　　　　　9. _____

10. The most common threading dies are the __?__ and __?__ types.　　　　　　10. _____
_____

11. A hand __?__ is a cutting tool used for finishing drilled or bored holes to an __?__ size and shape.　　　　　　11. _____
_____

12. A metal pin made of soft steel, brass, copper or aluminum with a round head on one end is called a __?__ .　　　　　　12. _____

Test #14 Threading and Reaming (Section 9, Unit 16)

Write the name of each tool or object in the proper space in the right-hand column.

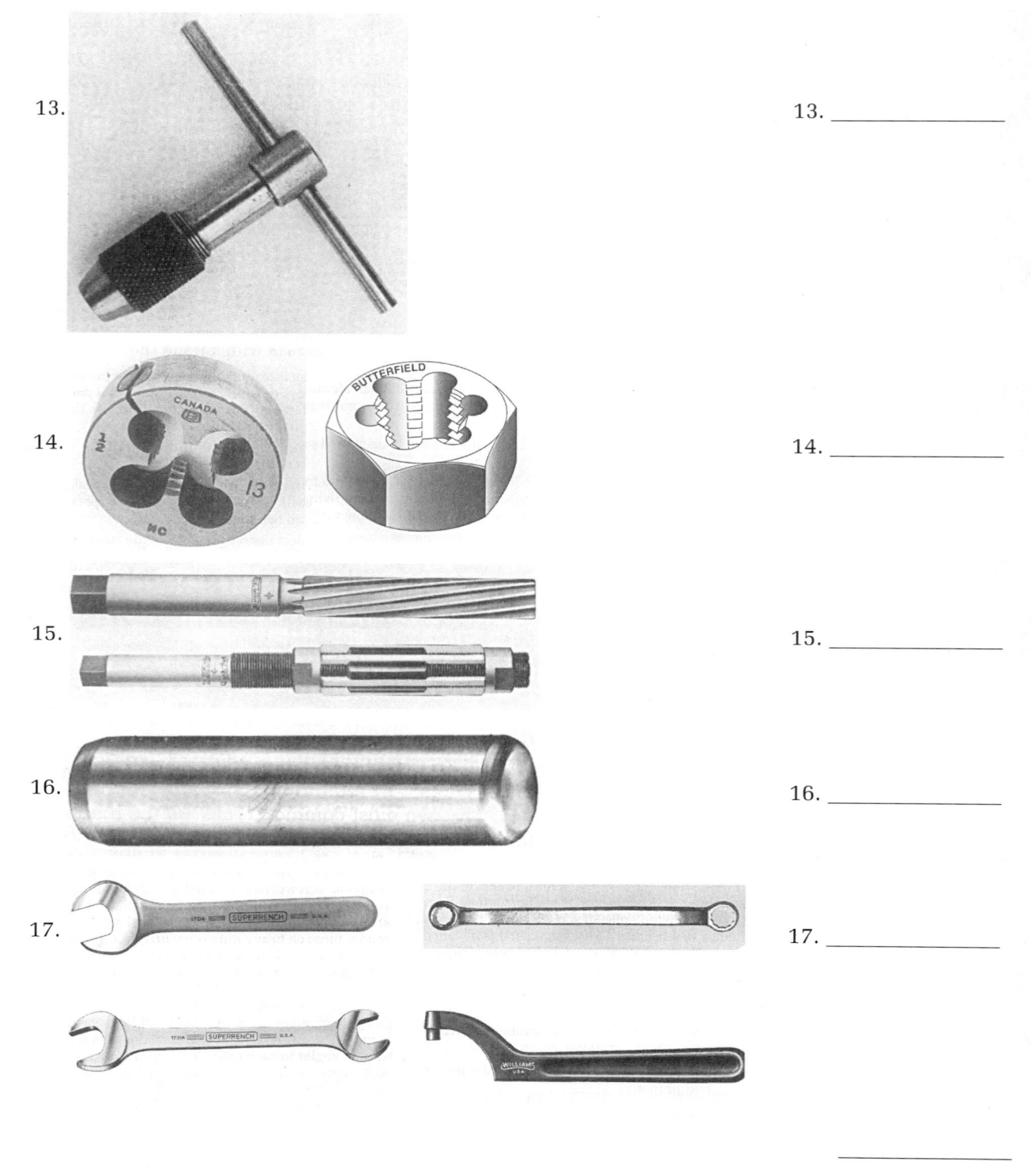

13. 13. _____

14. 14. _____

15. 15. _____

16. 16. _____

17. 17. _____

_____

Total—25 Marks

46   Lab Manual to Accompany Machine Tool and Manufacturing Technology

# REVIEW TEST #3
## BENCH AND HAND TOOLS (SECTION 9—UNITS 15 AND 16)

**PART 1**

Place the letter of each tool or operation beside the proper name in the right-hand column.

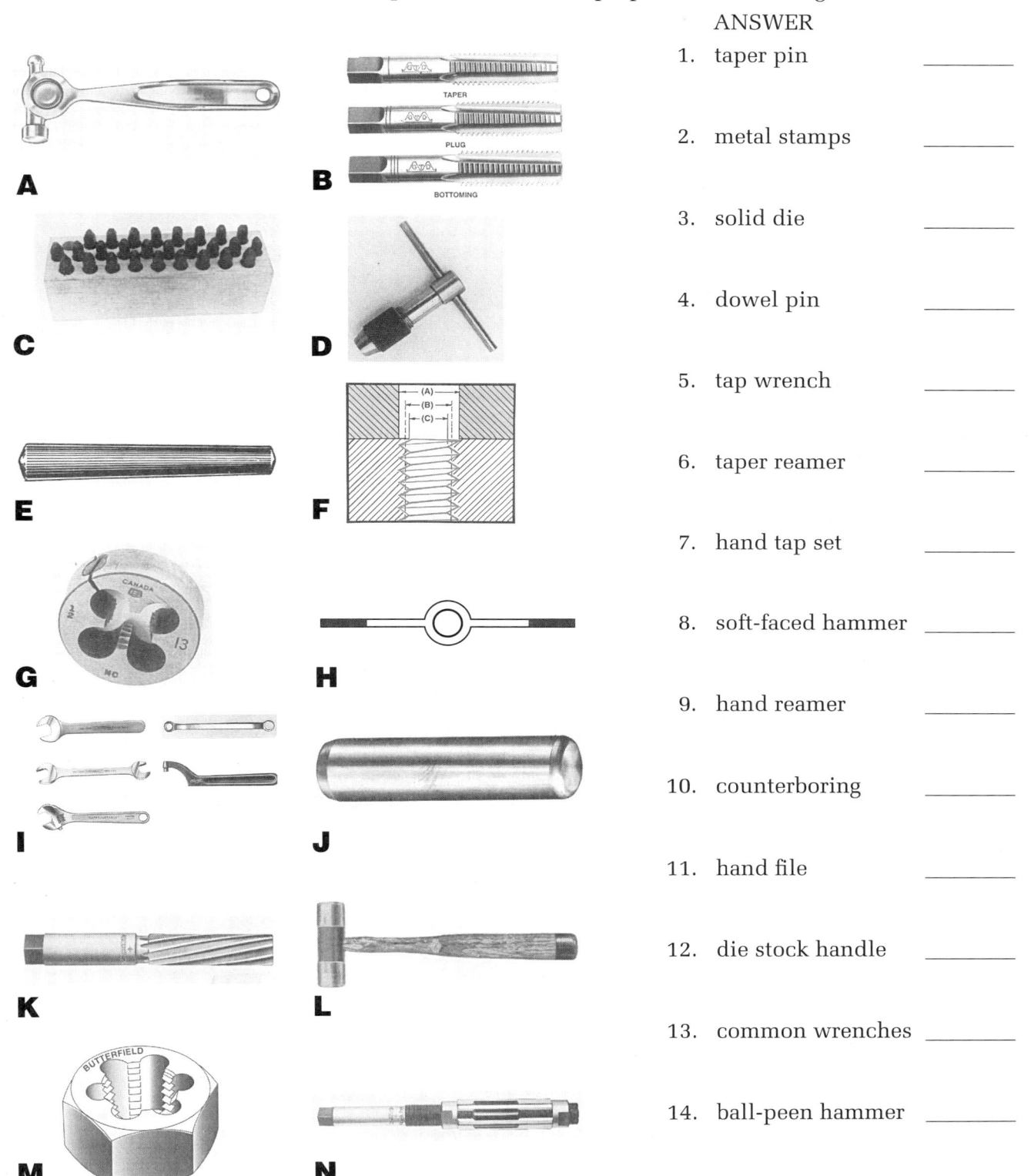

ANSWER

1. taper pin          _____
2. metal stamps       _____
3. solid die          _____
4. dowel pin          _____
5. tap wrench         _____
6. taper reamer       _____
7. hand tap set       _____
8. soft-faced hammer  _____
9. hand reamer        _____
10. counterboring     _____
11. hand file         _____
12. die stock handle  _____
13. common wrenches   _____
14. ball-peen hammer  _____

O

P

Q  R

S  T

U  V

W  X

Y  Z

15. machine screws  _____

16. bench vise  _____

17. expansion reamer  _____

18. round adjustable die  _____

19. self-tapping screws  _____

20. hand hacksaw  _____

21. common screwdrivers  _____

22. checking for squareness  _____

23. tightening hacksaw  _____

24. toolmaker's hammer  _____

25. T.D. cross-section  _____

26. file cross-section  _____

## PART 2

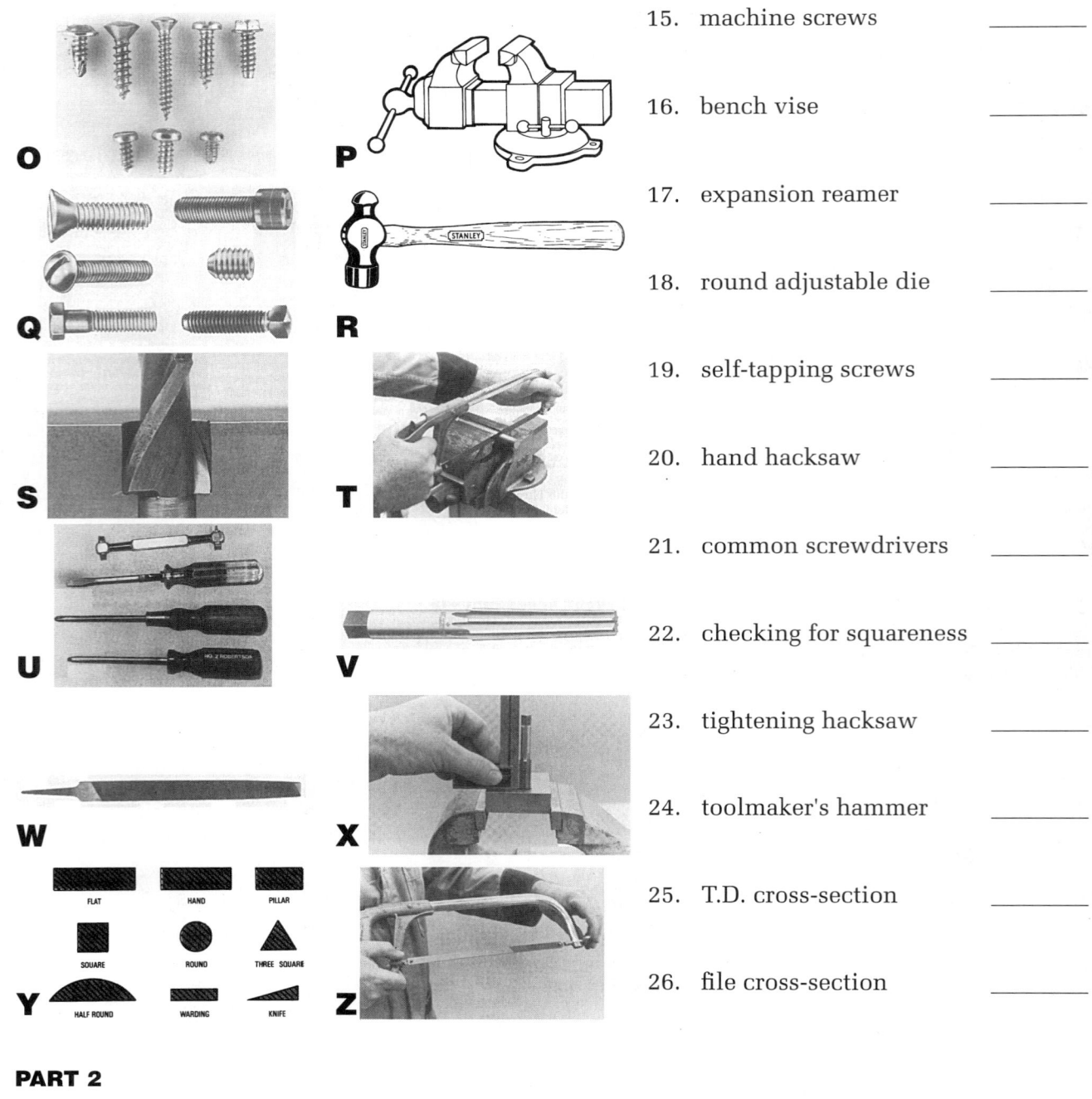

Place the correct word(s) in the space(s) at the right-hand side of the page to make the statement complete and true.

27. Hand tools fall into two categories: __?__ and __?__-__?__ tools.

28. The machinist vise is a __?__-__?__ device.

29. The toolmaker's hammer is also referred to as a __?__ .

30. For reasons of safety, never strike the __?__ of a hammer against __?__ .

27. _____
    _____-_____

28. _____-_____

29. _____

30. _____
    _____

48  Lab Manual to Accompany Machine Tool and Manufacturing Technology

31. The __?__ is a hand tool used to cut metal.  31. _____

32. The distance between each tooth on a saw blade is called the __?__.  32. _____

33. In mounting a blade on a hacksaw, the __?__ of the blade should point __?__ from the handle.  33. _____
_____

34. When cutting thin material, at least __?__ __?__ should bear on the work at all times.  34. _____
_____

**PART 3**

Consider whether each statement is true or false, then indicate your choice by circling the proper letter in the right-hand column.

| | | | |
|---|---|---|---|
| 35. | Both single and double-cut files are manufactured in various degrees of coarseness. | 35. | T  F |
| 36. | The reason a file should never be used without a handle is that the tang could easily puncture a hand or an arm. | 36. | T  F |
| 37. | The most common taps have five flutes cut length-wise across the threads. | 37. | T  F |
| 38. | Inch taps have the letter M stamped on the shank of the tap. | 38. | T  F |
| 39. | Tap drills should leave enough material in the hole to produce 75% of a full thread. | 39. | T  F |
| 40. | Because taps are hard and brittle, they are easily broken. | 40. | T  F |
| 41. | Cutting fluid is required when tapping brass or cast iron. | 41. | T  F |
| 42. | Solid dies can be turned onto the thread with an adjustable wrench. | 42. | T  F |
| 43. | A hand reamer is used for finishing holes to an accurate size and shape. | 43. | T  F |
| 44. | Hand reamers should never be used under mechanical power. | 44. | T  F |
| 45. | Turning a reamer backwards will damage the cutting edges. | 45. | T  F |
| 46. | Machine screws are never used for assembly work. | 46. | T  F |

_____

Total—50 marks

# TEST #15
# CUT-OFF SAWS (SECTION 10, UNIT 17)

Metal-cutting saws are used in manufacturing industries to cut rough lengths from bar stock before they are finished to size by machining. The two main types of cut-off saws are the horizontal and the vertical machines.

Place the correct word(s) in the space(s) at the right-hand side of the page to make the statement complete and true.

ANSWER

1. The power hacksaw is a __?__ type of saw.   1. _____

2. The saw __?__ and __?__ of a power hacksaw travel back and forth.   2. _____ _____

3. On the power hacksaw __?__ -pressure is applied automatically on the __?__ stroke.   3. _____ _____

4. The horizontal bandsaw has an endless blade which cuts __?__ in one __?__ .   4. _____ _____

5. The abrasive cut-off saw can be used __?__ , however, __?__ fluid is generally used to cool the saw and work.   5. _____ _____

6. Saw blades for the power hacksaw are usually hardened __?__ .   6. _____

7. Flexible blades used on bandsaws have only the __?__ hardened.   7. _____

8. When cutting large sections a __?__ -pitch blade provides chip clearance and helps to increase tooth __?__ .   8. _____ _____

9. In selecting a saw blade pitch, always make sure that __?__ teeth of the blade are in contact with the work at all times.   9. _____

10. When installing a blade, the teeth of the blade should be pointing in the direction of saw __?__ .   10. _____

11. When cutting, the blade __?__ should suit the type and thickness of the __?__ .   11. _____ _____

12. The vise of a horizontal bandsaw can be swiveled for making __?__ cuts.   12. _____

Test #15 Cut-Off Saws (Section 10, Unit 17)

13. Long work extending beyond the vise should be __?__ by a floor stand.

14. Never attempt to mount, measure, or remove work unless the saw is __?__ .

13. _____

14. _____

_____

Total—20 marks

# TEST #16
# CONTOUR-CUTTING BANDSAW (SECTION 10, UNIT 18)

The contour-cutting bandsaw is widely used by industry as a fast and economical method of cutting metal. It has provided industry with an economical and fairly accurate method of sawing, filing, and polishing straight and contour shapes.

Consider whether each statement is true or false, then indicate your choice by circling the proper letter in the right-hand column.

|     |     | ANSWER |
| --- | --- | --- |
| 1.  | Saw guides help to guide and support the vertical saw blade. | 1. T F |
| 2.  | The recommended cutting speeds for various materials and pitches of blades are listed on the job selector. | 2. T F |
| 3.  | Contour bandsaw blades are available in only one type of tooth form. | 3. T F |
| 4.  | Saw blades are available in only one type of tooth set. | 4. T F |
| 5.  | Contour bandsaws have a welding attachment to make a continuous sawband loop. | 5. T F |
| 6.  | It is necessary to anneal the saw band after welding. | 6. T F |
| 7.  | In mounting a saw blade, the teeth should be pointing away from the table. | 7. T F |
| 8.  | A work-holding jaw or a piece of wood should be used to feed the work into the saw. | 8. T F |
| 9.  | The upper pulley is used to support and tension the saw band. | 9. T F |
| 10. | The regular tooth blade allows rapid chip removal without increasing frictional heat. | 10. T F |
| 11. | The skip tooth blade has wide tooth spacing and is ideal for high-speed cutting of ferrous metals. | 11. T F |
| 12. | When cutting large work sections, a coarse pitch blade should be used. | 12. T F |
| 13. | To make angular cuts, the blade may be tilted up to 45°. | 13. T F |
| 14. | In calculating blade length, the center distance between pulleys and the circumference of one pulley is used. | 14. T F |
| 15. | As a general rule, not less than two teeth should be in cutting contact with the work at all times. | 15. T F |

16. A wide saw blade can be used to cut a small radius.  16.  T  F
17. When mounting a file band, the unhinged end of the file segment should be pointed downward.  17.  T  F
18. The best filing speeds are between 50 and 100 feet (15 and 30 m) per minute.  18.  T  F
19. Heavy work pressure to the file band produces a better finish and prevents clogging.  19.  T  F
20. The machine should be stopped before attempting to clean the band file.  20.  T  F

Place the name of each saw blade in the proper space in the right-hand column.

21. 21. _____

22. 22. _____

23. 23. _____

24. 24. _____

25. 25. _____

Total—25 marks

# TEST #17
# DRILL PRESS TYPES (SECTION 11, UNIT 19)

The drill press is perhaps one of the first mechanical devices developed by humans for the purpose of drilling holes. The main purpose of a drill press is to hold, revolve, and feed a twist drill to produce a hole in a piece of material.

Select the correct answer for each question and indicate your choice by circling the correct letter in the right-hand column.

ANSWER

1. The most common drill presses found in a machine shop are the
   (A) bench type
   (B) floor type
   (C) both of these
   (D) none of these

   1.　A　B　C　D

2. The size of any drill press is generally given as the distance from the edge of the column to the center of the drill press
   (A) table
   (B) spindle
   (C) base
   (D) drill chuck

   2.　A　B　C　D

3. The main parts of a bench and floor model drill press include
   (A) column
   (B) table
   (C) drilling head
   (D) all of these

   3.　A　B　C　D

4. The most common drill press toolholding device is a
   (A) drill chuck
   (B) drill sleeve
   (C) drill socket
   (D) spindle sleeve

   4.　A　B　C　D

5. The drill chucks which hold straight shank tools have
   (A) two jaws
   (B) three jaws
   (C) four jaws
   (D) six jaws

   5.　A　B　C　D

Consider whether each statement is true or false, then indicate your choice by circling the proper letter in the right-hand column.

6. Drill sleeves are used to adapt the cutting tool shank to the machine spindle if it is too small.

   6.　T　F

Test #17 Drill Press Types (Section 11, Unit 19)　　　55

7. A drill socket is used when the drill press spindle hole is too small for the tapered shank of the drill.   7.  T  F

8. When mounting a taper shank tool, the tang must first be aligned with the slot in the spindle hole.   8.  T  F

9. A drill drift is a square-shaped tool.   9.  T  F

10. When removing a taper shank tool, a wooden block should be placed on the drill press table.   10.  T  F

11. To avoid accidents, a chuck key should never be left in a drill chuck.   11.  T  F

12. As a drill breaks through the workpiece, drilling pressure should be increased slightly.   12.  T  F

Place the name of each tool or operation in the proper space in the right-hand column.

13.     14.     13. _____

14. _____

15.     16.     15. _____

16. _____

17.     18.     17. _____

18. _____

19.

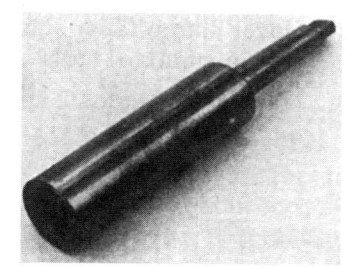

20.

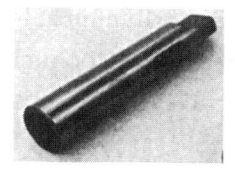

19. _____

20. _____


_____

Total—20 marks

Test #17 Drill Press Types (Section 11, Unit 19)  57

# TEST #18
# TWIST DRILLS (SECTION 11, UNIT 20)

The twist drill is one of the most common cutting tools found in a machine shop. As with any cutting tool, proper selection, care, and use will increase the life of a twist drill and produce high-quality work.

Place the correct word(s) in the space(s) at the righthand side of the page to make the statement complete and true.

ANSWER

1. A twist drill is an __?__-__?__ tool used to produce a hole in a piece of material.

    1. _____-_____

2. The most common drill has __?__ cutting edges, and __?__ straight or __?__ flutes.

    2. _____
       _____
       _____

3. The purpose of the flutes is to admit __?__ fluid, and allow the __?__ to escape during the drilling operation.

    3. _____
       _____

4. The __?__ edge has a negative __?__ and does not produce a good cutting action.

    4. _____
       _____

5. The most efficient cutting action occurs when the __?__ or cutting edges of the drill contact the metal.

    5. _____

6. The most common twist drills in a machine shop are made of __?__-__?__ steel and __?__ carbides.

    6. _____-_____
       _____

7. There are __?__ main sections on most twist drills.

    7. _____

8. The __?__ is a narrow, raised section on the body of the drill next to the __?__.

    8. _____
       _____

9. The __?__ on a drill gradually increases in __?__ toward the shank to give the drill __?__.

    9. _____
       _____
       _____

10. The average lip clearance on a twist drill is from __?__ to __?__, depending on the type of material to be drilled.

    10. _____
        _____

11. Inch drills are designated by __?__, __?__, and __?__ systems.

    11. _____
        _____
        _____

Test #18 Twist Drills (Section 11, Unit 20)

12. The size of a straight shank drill is marked on the __?__ .   12. _____

13. Taper shank drills are generally stamped on the neck between the __?__ and __?__ .   13. _____
    _____

_____

Total—25 marks

# TEST #19
# PRODUCING AND FINISHING HOLES (SECTION 11, UNIT 21)

For best results when drilling and finishing holes, always use the correct speeds and feeds to suit the drill diameter and the workpiece material. The use of cutting fluid during the drilling and finishing operations will prolong the tool life and increase productivity.

Consider whether each statement is true or false, then indicate your choice by circling the proper letter in the right-hand column.

|     |                                                                                                                                      |     | ANSWER |   |
|-----|--------------------------------------------------------------------------------------------------------------------------------------|-----|--------|---|
| 1.  | Heat and friction will dull a cutting drill quickly and produce a poor finish.                                                       | 1.  | T      | F |
| 2.  | The main purpose of cutting fluids is to prolong tool life and increase manufacturing productivity.                                  | 2.  | T      | F |
| 3.  | Synthetic cutting fluids are not as clean or efficient as cutting oil and soluble oils.                                              | 3.  | T      | F |
| 4.  | A center drill is a tool that combines the operation of drilling and countersinking.                                                 | 4.  | T      | F |
| 5.  | Center drills are available in two types: plain and regular.                                                                         | 5.  | T      | F |
| 6.  | A regular center drill is designed to protect the top edge of a bearing surface from becoming damaged.                               | 6.  | T      | F |
| 7.  | It is considered good practice to first spot a center punch mark with a center drill.                                                | 7.  | T      | F |
| 8.  | The most common method of holding small workpieces is by means of a vise.                                                            | 8.  | T      | F |
| 9.  | When work is clamped to the table, parallels should be placed between the table and the work to avoid drilling into the table.       | 9.  | T      | F |
| 10. | The web of a drill increases as the size of a drill increases.                                                                       | 10. | T      | F |
| 11. | To relieve the drilling pressure of a large drill, a pilot hole should first be drilled to act as a guide for the drill to follow.   | 11. | T      | F |
| 12. | The size of a pilot hole drilled should be slightly smaller than the web thickness of the drill to be used.                          | 12. | T      | F |
| 13. | Larger drills tend to create their own paths and will not follow the pilot hole.                                                     | 13. | T      | F |

14. When drilling round work, it may be supported by a V-block in a drill vise, or it may be mounted on V-blocks and clamped to the table.  14.  T  F

15. The speed for countersinking is generally one-half of the recommended drilling speed.  15.  T  F

16. The speed for counterboring is generally one-half of the recommended drilling speed.  16.  T  F

17. The speed for reaming is generally one-quarter of the recommended drilling speed.  17.  T  F

18. The speed for boring is generally one-half of that used for drilling a hole of the same size.  18.  T  F

19. When spot-facing, the drill press should be set to approximately one-quarter of the drilling speed.  19.  T  F

20. Tapping in a drill press can be performed either by hand or with a tapping attachment.  20.  T  F

Total—20 marks

# REVIEW TEST #4
# DRILL PRESS (SECTION 11—UNITS 17 TO 21)

## PART 1

Place the letter of each drill press or part, cutting tool, accessory, or operation beside the proper name in the right-hand column.

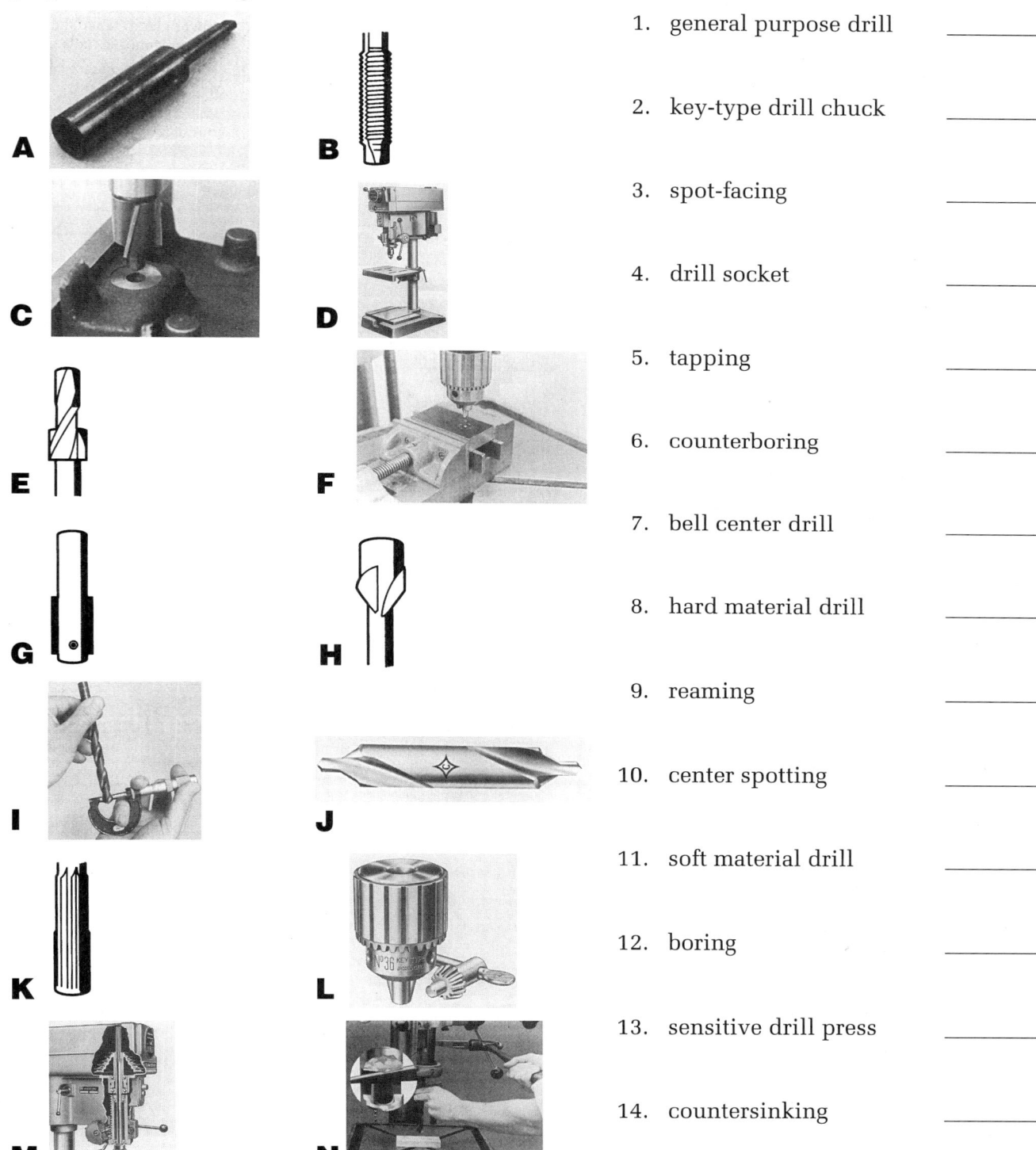

1. general purpose drill  _____

2. key-type drill chuck  _____

3. spot-facing  _____

4. drill socket  _____

5. tapping  _____

6. counterboring  _____

7. bell center drill  _____

8. hard material drill  _____

9. reaming  _____

10. center spotting  _____

11. soft material drill  _____

12. boring  _____

13. sensitive drill press  _____

14. countersinking  _____

Review Test #4 Drill Press (Section 11—Units 17 to 21)

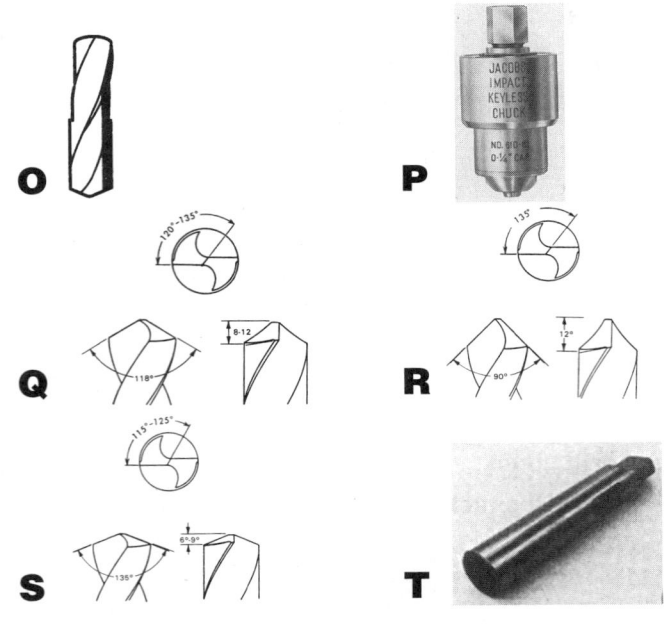

| | | |
|---|---|---|
| 15. | drilling head | _____ |
| 16. | drill drift | _____ |
| 17. | drill sleeve | _____ |
| 18. | drill measuring | _____ |
| 19. | keyless chuck | _____ |
| 20. | drilling | _____ |

**PART 2**

Place the name of each illustrated twist drill part in the proper space in the right-hand column.

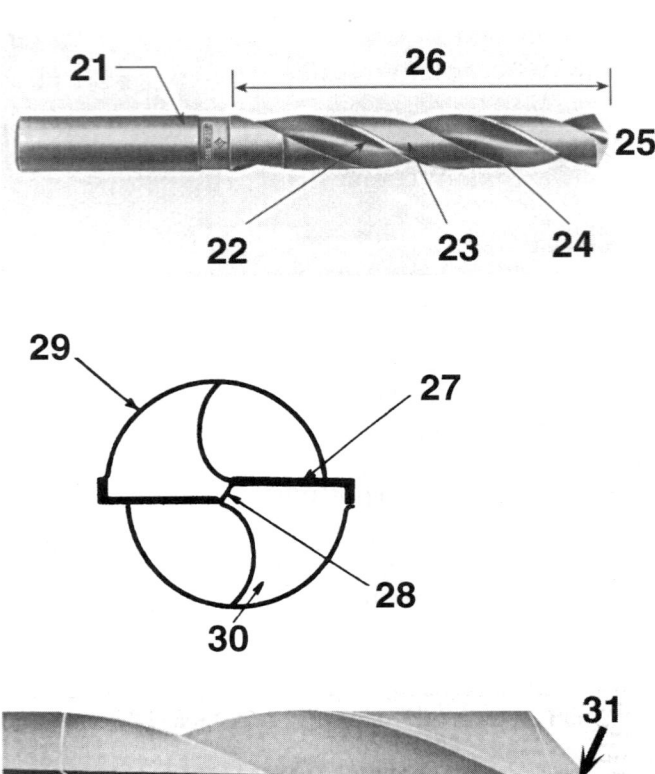

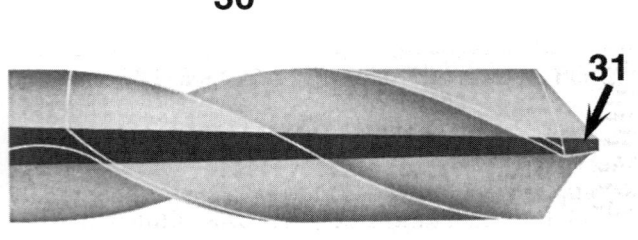

ANSWER

21. _____
22. _____
23. _____
24. _____
25. _____
26. _____
27. _____
28. _____
29. _____
30. _____
31. _____

64  Lab Manual to Accompany Machine Tool and Manufacturing Technology

**PART 3**

Place the correct word(s) in the space(s) at the right-hand side of the page to make the statement complete and true.

32. The size of any drill press is generally given as the distance from the edge of the __?__ to the center of the drill press __?__ .

    32. _____
        _____

33. To prevent accidents, never wear __?__ clothing around machinery.

    33. _____

34. Always wear __?__ glasses when operating any __?__ .

    34. _____
        _____

35. Twist drill sizes are designated by __?__ , __?__ and __?__ systems.

    35. _____
        _____
        _____

36. Metric drills are always measured in __?__ sizes.

    36. _____

37. The speed of a drill is measured in surface __?__ per minute or in surface __?__ per minute.

    37. _____
        _____

38. For best results in producing a hole, it requires the use of proper __?__ and __?__ to suit the workpiece __?__ and the size of the __?__ .

    38. _____
        _____
        _____
        _____

39. Cutting fluids are used to __?__ tool life and increase __?__ .

    39. _____
        _____

40. To prevent a drill from jamming as it breaks __?__ the work, ease up on the drilling __?__ .

    40. _____
        _____

_____

Total—50 marks

# TEST #20
# LATHE TYPES AND CONSTRUCTION ( SECTION 12, UNIT 22 )

The main function of the engine lathe is to machine cylindrical forms on round work. Over the years, the basic lathe led to the development of other machines such as the turret lathe, screw machine, boring mill, and the more advanced computer numerical control chucking and turning centers.

Place the correct word(s) in the space(s) at the right-hand side of the page to make the statement complete and true.

ANSWER

1. The main function of the engine lathe is to produce cylindrical forms by using a __?__ cutting tool which is brought into contact with a __?__ workpiece.

   1. _____
      _____

2. The accuracy of the work produced on a lathe is controlled by the skills of the machine __?__ .

   2. _____

3. A lathe equipped with digital readout helps to improve the __?__ and __?__ of any lathe.

   3. _____
      _____

4. The size of a lathe is determined by the largest __?__ of work which may be revolved over the __?__ , and the longest work held between __?__ .

   4. _____
      _____
      _____

5. Most modern lathes are __?__-__?__ which allows them to be set for various __?__ speeds quickly.

   5. _____-_____
      _____

6. The quick-change gearbox contains a number of different sized gears which control the various speeds for turning and __?__-__?__ operations.

   6. _____-_____

7. The carriage consists of three main parts, the __?__ , the __?__ and the cross-slide.

   7. _____
      _____

8. The compound rest can be swiveled to any angle for __?__-__?__ operations.

   8. _____-_____

9. Adjustment screws can be used for __?__ the tailstock for taper-turning.

   9. _____

10. Graduated micrometer collars may indicate the distance a cutting tool has been moved __?__ the work or the amount which will be removed from the __?__ .

    10. _____
        _____

Test #20 Lathe Types and Construction ( Section 12, Unit 22 )

11. Always __?__ the lathe before making any measurement of any kind.

12. __?__ cuts on long slender pieces can cause the work to __?__ and fly out of the machine.

11. _____

12. _____
   _____

_____

Total—20 marks

# TEST #21
# LATHE ACCESSORIES AND TOOLING (SECTION 12, UNIT 23)

The conventional lathe becomes more flexible with accessories and tooling which fit the spindle, carriage, compound rest, and tailstock. These accessories enable an operator to produce parts much faster and more accurately.

Select the correct answer for each question and indicate your choice by circling the correct letter in the right-hand column.

ANSWER

1. Work-holding devices that are used to hold and drive work on a lathe include
   (A) drive plate
   (B) chucks
   (C) lathe centers
   (D) all of these

   1.  A   B   C   D

2. Which type of spindle nose is used on an engine lathe?
   (A) tapered
   (B) cam-lock
   (C) threaded
   (D) all of these

   2.  A   B   C   D

3. The revolving tailstock center is used to replace
   (A) live center
   (B) dead center
   (C) collet center
   (D) drive center

   3.  A   B   C   D

4. Which of the following statements does not apply to a universal chuck?
   (A) all jaws move together
   (B) holds octagonal shapes
   (C) holds hexagonal shapes
   (D) holds round shapes

   4.  A   B   C   D

5. Which of the following statements does not apply to a four-jaw chuck?
   (A) holds round shapes
   (B) holds square shapes
   (C) holds hexagonal shapes
   (D) holds octagonal shapes

   5.  A   B   C   D

Place the correct word(s) in the space(s) at the right-hand side of the page to make the statement complete and true.

ANSWER

6. A __?__ chuck has the mechanical features of both independent and universal chucks.

6. _____

7. The __?__ chuck is a draw-in __?__ or collet that fits into the __?__ spindle.

7. _____
_____
_____

8. The __?__ collet chuck uses an impact-tightening __?__ to open and close the collet on the workpiece.

8. _____
_____

9. The most common accessories mounted on the lathe carriage or bed are the __?__ rest and the __?__ rest.

9. _____
_____

10. To machine metal in a lathe, a cutting tool called a __?__ is used.

10. _____

11. For general turning, a __?__ toolholder is used.

11. _____

12. When turning work toward the headstock, a __?__-__?__ toolholder is used.

12. _____-_____

13. The standard or __?__ toolpost is generally supplied with the lathe.

13. _____

14. Turret-type toolposts are designed to hold __?__ cutting tools, which can be __?__ into various positions.

14. _____
_____

15. A __?__ taper hole in the tailstock spindle allows it to hold cutting tools and accessories.

15. _____

Place the name of each accessory or operation in the proper space in the right-hand column.

16.

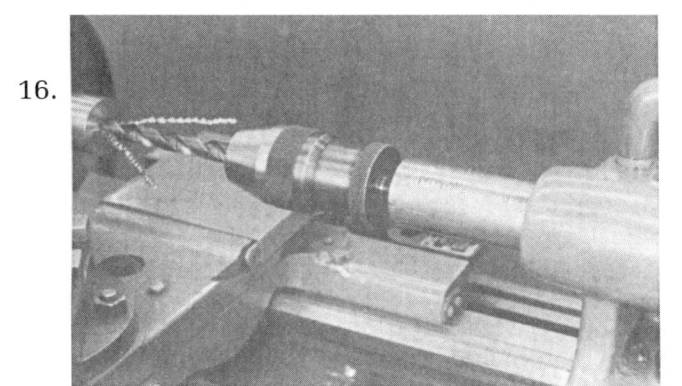

16. _____

17.

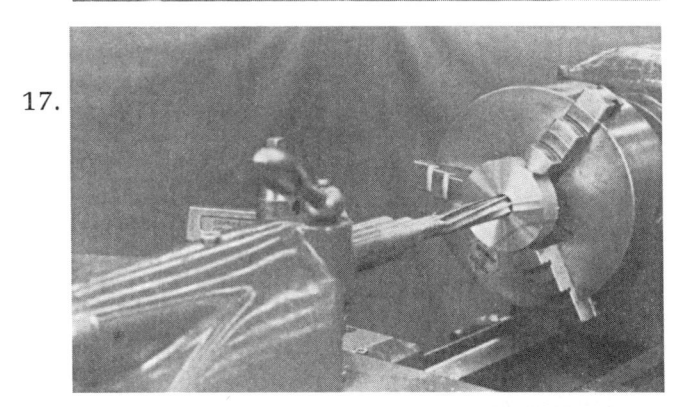

17. _____

18.

18. _____

19.

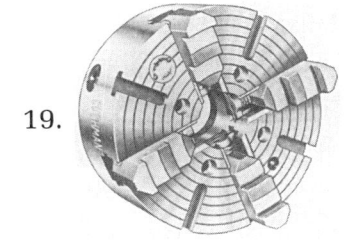

19. _____-_____ _____

20.

20. _____-_____ _____

_____
Total—25 marks

Test #21 Lathe Accessories and Tooling (Section 12, Unit 23)

71

# TEST #22
# CUTTING SPEEDS AND FEEDS (SECTION 12, UNIT 24)

In order to machine metal efficiently, it is important to understand how diameter, work material and cutting tool affect the speeds and feeds of a lathe. Using the correct speeds and feeds will prolong tool life and increase productivity; incorrect speeds and feeds will increase tool wear and result in unnecessary machine downtime.

Consider whether each statement is true or false, then indicate your choice by circling the proper letter in the right-hand column.

|     |                                                                                                                              |     | ANSWER |     |
| --- | ---------------------------------------------------------------------------------------------------------------------------- | --- | ------ | --- |
| 1.  | The only factors which affect the lathe speed are the work diameter and the work material.                                   | 1.  | T      | F   |
| 2.  | Cutting speed is the rate at which a point on the work circumference passes the cutting tool in one minute.                  | 2.  | T      | F   |
| 3.  | Cutting speed is expressed as feet or meters per minute.                                                                     | 3.  | T      | F   |
| 4.  | Cutting speeds have been recommended by cutting tool and metal manufacturers.                                                | 4.  | T      | F   |
| 5.  | If the cutting speed is increased for the workpiece, the result will be increased productivity and longer tool life.         | 5.  | T      | F   |
| 6.  | The spindle speed and cutting speed both mean the same thing.                                                                | 6.  | T      | F   |
| 7.  | Spindle speeds are measured in revolutions per minute.                                                                       | 7.  | T      | F   |
| 8.  | If a lathe cannot be set to the calculated spindle speed, it should be set at the next higher speed.                         | 8.  | T      | F   |
| 9.  | On a geared-head lathe, speeds may be changed while the lathe is running.                                                    | 9.  | T      | F   |
| 10. | The r/min. for any workpiece can be calculated by using the formula: $\dfrac{CS \times D}{4}$                                | 10. | T      | F   |
| 11. | Feed may be defined as the distance the cutting tool advances along the work for every revolution.                           | 11. | T      | F   |
| 12. | The feed is dependent on the rate at which the leadscrew or feed rod turns.                                                  | 12. | T      | F   |
| 13. | As a general rule, only three cuts should be taken to bring a diameter to finish size.                                       | 13. | T      | F   |

14. The purpose of a roughing cut is to remove excess material gradually.     14. T   F

15. In order to machine work in the shortest possible time, spindle speeds and feeds should always be increased above recommended speeds and feeds.     15. T   F

16. When rough turning, the condition of the lathe, workpiece rigidity, and the type of cutting tool will determine the depth of cut.     16. T   F

17. For general purpose rough machining, a feed of not more than .005 in. (0.1 mm) should be used.     17. T   F

18. In most cases, the depth of the finish cut should be less than .002 in. (0.04 mm).     18. T   F

19. The width or depth of the chip is determined by the depth of cut taken by the cutting tool.     19. T   F

20. To calculate machining time, factors such as speed, feed, and depth of cut must be considered.     20. T   F

Total—20 marks

# TEST #23
# MOUNT AND REMOVE ACCESSORIES (SECTION 12, UNIT 25)

Mounting and removing lathe accessories is an important function in any metalworking shop. It is very important that this operation be carried out safely and in proper sequence to have the accessories seated properly and to preserve lathe accuracy.

Place the correct word(s) in the space(s) at the right-hand side of the page to make the statement complete and true.

ANSWER

1. Lathe centers are removed from a lathe for __?__ , obtaining center __?__ , and to replace __?__ centers.

2. Along with dirt, __?__ or __?__ will prevent centers from being properly seated.

3. A center should be inserted into the headstock spindle with a sharp __?__ .

4. Headstock centers are removed from the spindle by using a __?__ bar.

5. To produce a __?__ diameter on work mounted between centers, the headstock and tailstock centers must be in __?__ .

6. The lathe tailstock consists of two halves, the __?__ and __?__ .

7. There are __?__ common methods used to check the __?__ of lathe centers.

8. The most accurate method of checking the alignment of lathe centers is to use a __?__ and a dial __?__ .

9. Lathe chucks are removed from a taper spindle nose by using a proper __?__-__?__ wrench.

10. Before mounting any chuck or spindle accessory, be sure that the electrical power switch is __?__ __?__ .

11. Work held in a three-jaw chuck should not extend more than __?__ times its __?__ beyond the chuck jaws.

Total—20 marks

# TEST #24
# MOUNT WORK BETWEEN CENTERS (SECTION 12, UNIT 26)

Much of the work in training programs is machined between lathe centers because it is possible to machine the work, remove it from the lathe, and replace it any number of times to the same accuracy.

Select the correct answer for each question and indicate your choice by circling the correct letter in the right-hand column.

|     |     |     | ANSWER |
| --- | --- | --- | --- |

1. For best rigidity, the toolholder's setscrew should be away from the toolpost about

    (A) a finger width  (C) a thumb width
    (B) a rule width    (D) one inch

    1.  A  B  C  D

2. The toolpost should be set in the compound rest at the

    (A) right-hand side  (C) middle
    (B) left-hand side   (D) end

    2.  A  B  C  D

3. The toolbit should never extend beyond the toolholder more than

    (A) 1/4 in. (6 mm)  (C) 3/8 in. (9 mm)
    (B) 1/8 in. (3 mm)  (D) 1/2 in. (12 mm)

    3.  A  B  C  D

4. For best results, work should be mounted in a lathe using the following centers:

    (A) two live     (C) two revolving
    (B) two dead     (D) one live, one revolving

    4.  A  B  C  D

5. A revolving tailstock center eliminates the need for

    (A) lubrication  (C) setting
    (B) friction     (D) heat

    5.  A  B  C  D

6. When facing between centers, the lathe centers must be

    (A) dead   (C) in-line
    (B) live   (D) rotating

    6.  A  B  C  D

7. When facing, the tool should be extended beyond the toolholder about          7.   A   B   C   D

   (A) 1/4 in. (6 mm)          (C) 1/2 in. (12 mm)
   (B) 3/8 in. (9 mm)          (D) 5/8 in. (15 mm)

8. When facing the entire end, what kind of tailstock center is recommended?          8.   A   B   C   D

   (A) quarter          (C) full
   (B) half             (D) three quarter

9. How may a workpiece be marked for length before facing?          9.   A   B   C   D

   (A) scriber          (C) pencil
   (B) center punch     (D) file mark

10. The facing tool should be fed from          10.   A   B   C   D

    (A) outside to center     (C) right to left
    (B) center to outside     (D) left to right

Total—10 marks

# TEST #25
# MACHINING BETWEEN CENTERS (SECTION 12, UNIT 27)

When machining between lathe centers, work can be removed and replaced more quickly than by any other work-holding device without disturbing machining accuracy.

Place the correct word(s) in the space(s) at the right-hand side of the page to make the statement complete and true.

ANSWER

1. Work that must be cut to the same diameter along the entire length involves the operation of __?__ turning.

   1. _____

2. On some lathes, the __?__ __?__ collar indicates the distance the cutting tool moves toward the workpiece.

   2. _____
   _____

3. As a general rule, a roughing cut should not reduce the diameter more than __?__ .

   3. _____

4. The purpose of a finish cut is to bring the workpiece to the __?__ size and provide a good __?__ finish.

   4. _____
   _____

5. The purpose of a trial cut is to set the __?__ tool to the diameter.

   5. _____

6. The purpose of a trial cut is to produce a __?__ diameter on the work.

   6. _____

7. The purpose of a trial cut is to also set the __?__ graduated collar to the __?__ .

   7. _____
   _____

Consider whether each statement is true or false, then indicate your choice by circling the proper letter in the right-hand column.

8. Whenever possible, a diameter should be measured with a micrometer.

   8. T    F

9. A lathe should always be stopped when measuring work.

   9. T    F

10. A rough-cut diameter should be .030 to .050 in. (0.8 to 1.3 mm) over the finished size required.

    10. T    F

11. It is considered good practice to bring a diameter to size by some filing.

    11. T    F

12. It is considered good practice to cover the lathe bed with a cloth when filing in a lathe.

    12. T    F

13. When filing, never use a file without a properly fitted handle.　　13.　T　F

14. To reduce pinning, a file should be rubbed with chalk after is has been cleaned.　　14.　T　F

15. Polishing is a finishing operation which is done to improve the surface finish.　　15.　T　F

16. Aluminum oxide abrasive cloth should be used for polishing nonferrous metals.　　16.　T　F

17. Silicon carbide abrasive cloth should be used to polish ferrous metals.　　17.　T　F

_____

Total—20 marks

# TEST #26
# KNURLING, GROOVING, SHOULDER TURNING
# (SECTION 12, UNIT 28)

Knurling, grooving and shoulder turning are operations which can be performed on work held between centers or in a chuck. These operations can improve appearance, increase strength, or provide clearance for mating parts.

Consider whether each statement is true or false, then indicate your choice by circling the proper letter in the right-hand column.

|     |                                                                                                               |     | ANSWER |   |
| --- | ------------------------------------------------------------------------------------------------------------- | --- | ------ | - |
| 1.  | Knurling is a process of impressing circular patterns on the workpiece.                                       | 1.  | T      | F |
| 2.  | Knurling tools for CNC machines have the knurling rolls set at a 30° angle.                                   | 2.  | T      | F |
| 3.  | Knurling is performed by setting the lathe at one-quarter the speed for turning.                              | 3.  | T      | F |
| 4.  | The center of the floating head of the knurling tool is set even with the dead center point.                  | 4.  | T      | F |
| 5.  | The knurling tool should be set at 80° to the work.                                                           | 5.  | T      | F |
| 6.  | A set of worn knurling rolls can produce an incorrect pattern.                                                | 6.  | T      | F |
| 7.  | The automatic carriage feed may be disengaged at any time without damaging the knurling pattern.              | 7.  | T      | F |
| 8.  | Grooving can also be performed with a knurling tool.                                                          | 8.  | T      | F |
| 9.  | The lathe should be set to one-quarter the turning speed for grooving.                                        | 9.  | T      | F |
| 10. | The grooving tool should be set slightly below center and at 90° to the work.                                 | 10. | T      | F |
| 11. | Work should be grooved to the proper depth using a steady crossfeed rate while applying cutting fluid.        | 11. | T      | F |
| 12. | If chatter develops during grooving, increase the speed.                                                      | 12. | T      | F |
| 13. | Safety goggles should be worn during a grooving operation.                                                    | 13. | T      | F |
| 14. | The terms shoulder and step have the same meaning.                                                            | 14. | T      | F |
| 15. | To machine a square shoulder to length, mark the correct length with a center punch or a light groove.        | 15. | T      | F |

16. For a filleted shoulder, the small diameter should be machined to the correct length minus two times the radius to be cut.     16.   T   F

17. When machining a filleted shoulder, the lathe should be set to the turning speed.     17.   T   F

18. A radius toolbit is used to produce the correct filleted shoulder form.     18.   T   F

19. Short angular shoulders can be cut using the side of a toolbit.     19.   T   F

20. Long angular shoulders can be machined by using the compound rest set to the required angle.     20.   T   F

Total—20 marks

# TEST #27
# TAPER TURNING (SECTION 12, UNIT 29)

Tapers are widely used in the mechanical trades to quickly and accurately align mechanical parts. Every machinist should know how to machine and fit tapers for a wide variety of uses in a machine shop.

Place the correct word(s) in the space(s) at the right-hand side of the page to make the statement complete and true.

ANSWER

1. A taper may be defined as a uniform __?__ or __?__ in __?__ measured along the work length.

1. _____
   _____
   _____

2. The American Standards Association classifies tapers used on machines and tools as self-__?__ and self-__?__ tapers.

2. _____
   _____

3. The __?__ taper is the standard inch taper used for twist drills, reamers, end mills and lathe center __?__ .

3. _____
   _____

4. Metric tapers are expressed as a ratio of one __?__ per unit of length.

4. _____

5. The main parts of a taper are the __?__ of taper, taper __?__ , large diameter, and small diameter.

5. _____
   _____

6. Inch tapers are expressed in taper per __?__ , taper per __?__ , or __?__ .

6. _____
   _____
   _____

7. Tapers may be cut on a lathe using any of three methods: offsetting the __?__ , using the taper __?__ , and by feeding the __?__ rest.

7. _____
   _____
   _____

8. For the best accuracy the cutting tool must be set on __?__ when machining a taper.

8. _____

9. Taper attachments can be used to cut __?__ and __?__ tapers on workpieces.

9. _____
   _____

10. When a __?__ taper attachment is used, the depth of cut can be set by the __?__ handle.

10. _____
    _____

11. The compound rest is used to cut short, __?__ tapers that are given in __?__ .

11. _____
    _____

Test #27 Taper Turning (Section 12, Unit 29)

12. External tapers can be checked for accuracy by using a taper  12. _____
    __?__ gage, a standard micrometer, or a special __?__        _____
    micrometer.

                                                                 _____
                                                                 Total—25 marks

# TEST #28
# MACHINING IN A CHUCK (SECTION 12, UNIT 30)

The operations for machining work held in a chuck are basically the same as for machining between centers. The work should be held short to assure rigidity, and long work should be supported by some means.

Place the correct word(s) in the space(s) at the right-hand side of the page to make the statement complete and true.

ANSWER

1. All work should be machined in __?__ setup before removing it from a three- or four-jaw chuck to ensure __?__ .

2. Chuck jaws should be tightened around the most __?__ part of the workpiece to prevent __?__ .

3. Long work should be supported by a __?__ tailstock center or a __?__ rest.

4. The toolpost should be positioned on the __?__ side of the compound rest.

5. Before starting the machine, the chuck should be __?__ one turn by __?__ to see that the jaws clear all lathe parts.

6. Three-jaw chucks are supplied with two sets of jaws: a __?__ and a __?__ set.

7. The __?__ jaws of a three-jaw chuck can be used to grip internal diameters.

8. The jaws on a four-jaw independent chuck are __?__ to permit a wide __?__ of work to be gripped.

9. Before attempting any finish turning operation, be sure that the work has returned to __?__ __?__ .

10. Cut-off tools are often called __?__ tools.

11. Before using a cut-off tool, the lathe carriage should be __?__ in position.

12. Internal operations such as __?__ , __?__ , and __?__ can be performed on work held in a lathe chuck.

13. Straight __?__ drills are generally held in a drill chuck mounted in the tailstock.

Test #28 Machining in a Chuck (Section 12, Unit 30)  85

14. For reaming in a lathe, set the spindle speed to __?__-__?__ the drilling speed.

14. _____-_____

15. When tapping in a lathe, a __?__ tap is preferred because it clears chips easily.

15. _____

16. For tapping in the lathe, lock the lathe spindle and turn the tap by __?__ .

16. _____

_____

Total—25 marks

# TEST #29
# THREADS AND THREAD CUTTING (SECTION 12, UNIT 31)

Screw threads are widely used as fastening devices, to transmit power, to move materials, and to control movement. Every machinist should be able to cut threads on an engine lathe.

Select the correct answer for each question and indicate your choice by circling the correct letter in the right-hand column.

ANSWER

1. A thread may be defined as a helical ridge of uniform sections found on a
   (A) circle
   (B) square
   (C) cylinder
   (D) hexagon

   1.  A  B  C  D

2. The quick-change gearbox is designed to quickly set the lathe for the required thread
   (A) pitch diameter
   (B) root diameter
   (C) crest
   (D) pitch

   2.  A  B  C  D

3. The thread-chasing dial shows the operator when to engage
   (A) carriage
   (B) split-nut lever
   (C) crossfeed lever
   (D) compound rest

   3.  A  B  C  D

4. The thread-chasing dial is graduated into the following divisions:
   (A) four
   (B) six
   (C) eight
   (D) twelve

   4.  A  B  C  D

5. In comparison to turning speed, the speed for thread cutting should be
   (A) the same
   (B) one-half
   (C) one-quarter
   (D) one-eighth

   5.  A  B  C  D

6. When cutting a 60° thread, the compound rest should be set at
   (A) 60°
   (B) 29°
   (C) 30°
   (D) 45°

   6.  A  B  C  D

Test #29 Threads and Thread Cutting (Section 12, Unit 31)

7. The included angle of a toolbit to cut an American National form thread is

(A) 60°  (C) 30°
(B) 29°  (D) 45°

7. A B C D

8. The threading toolbit should be set even with the center point and against the work at

(A) 60°  (C) 90°
(B) 45°  (D) 30°

8. A B C D

9. Before cutting a thread, the end of the work should be

(A) flattened  (C) filed
(B) chamfered  (D) filleted

9. A B C D

10. General-purpose threads can be checked with a

(A) master screw  (C) master nut
(B) master gage   (D) master bolt

10. A B C D

Place the correct word(s) in the space(s) at the right-hand side of the page to make the statement complete and true.

11. Quick-change gear lathes may be changed for cutting metric threads by using two change gears having __?__ and __?__ teeth.

11. _____
    _____

12. The relationship of the inch and the metric system of measurement becomes clearer when it is understood that __?__ inch is equal to __?__ centimeters.

12. _____
    _____

13. Thread spacing is measured with a metric __?__ __?__ gage.

13. _____
    _____

14. When unfinished threaded work has been removed and replaced, the __?__ tool has to be __?__ .

14. _____
    _____

15. To check threads for depth, angle, and accuracy, thread __?__ gages and thread __?__ are commonly used.

15. _____
    _____

Total—20 marks

# REVIEW TEST #5
# LATHES, ACCESSORIES, AND THREAD CUTTING
# (UNITS 22 TO 31)

## PART 1

Place the letter of each lathe part beside the proper name in the right-hand column.

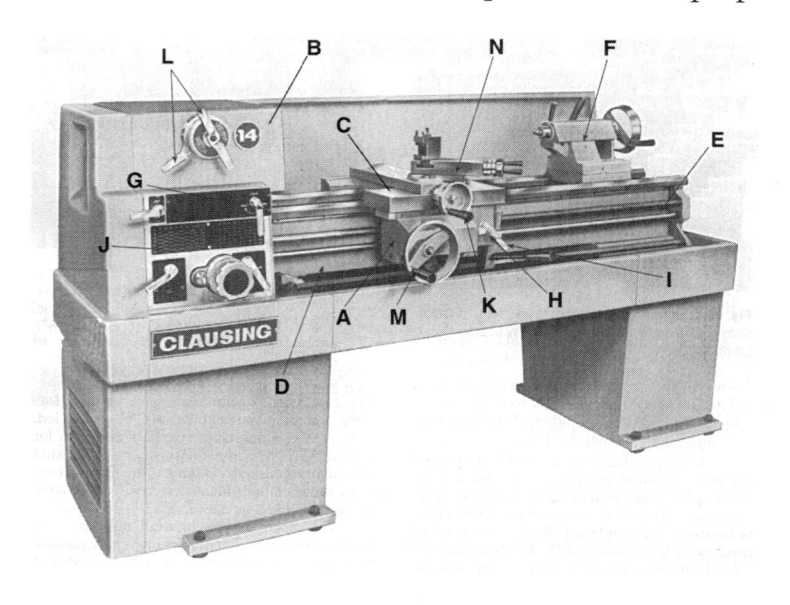

1. tailstock  _____
2. leadscrew  _____
3. automatic feed lever  _____
4. apron  _____
5. headstock  _____
6. quick-change gearbox  _____
7. split-nut lever  _____
8. saddle  _____
9. feed reverse lever  _____
10. bed  _____
11. compound rest  _____
12. crossfeed handle  _____
13. apron handwheel  _____
14. spindle speed levers  _____

## PART 2

Place the letter of each thread profile beside the proper name in the right-hand column.

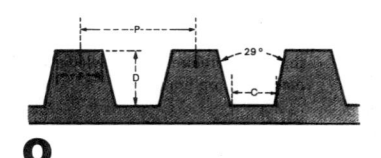

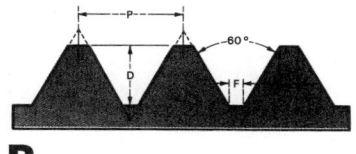

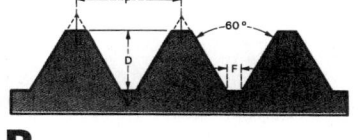

15. square thread  _____

16. International metric  _____

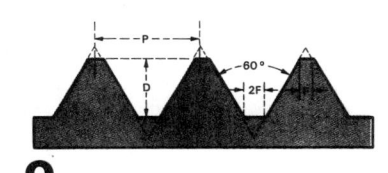

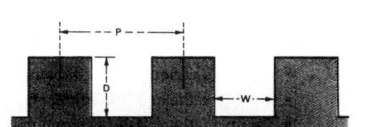

17. American National Form  _____

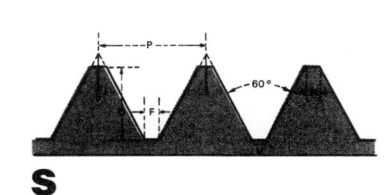

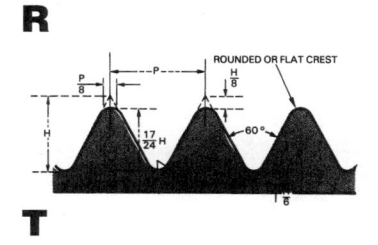

18. American National Acme  _____

19. Unified screw  _____

20. ISO metric thread  _____

# PART 3

Place the name of each lathe accessory or operation in the right-hand column.

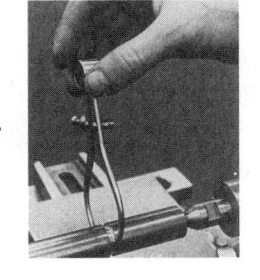

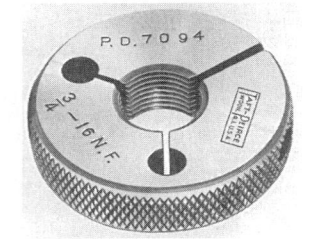

21.

22.

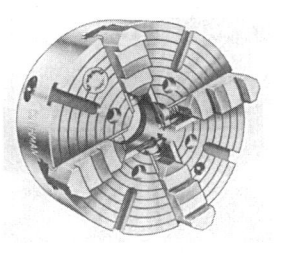

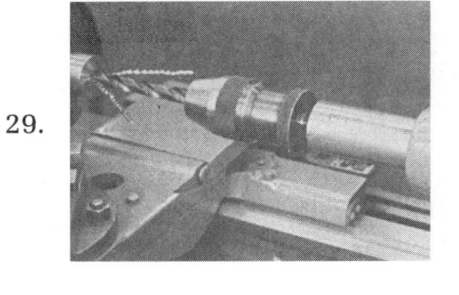

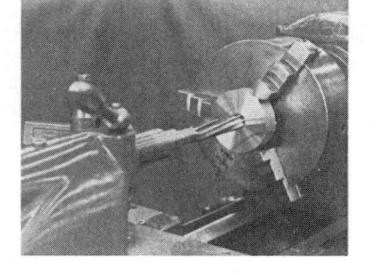

21. _____

22. _____

23.

24.

23. _____

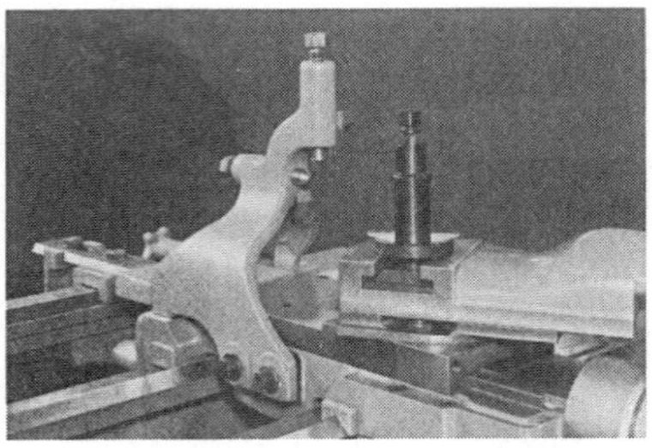

24. _____

25.

26.

25. _____

26. _____

27.

28.

27. _____

28. _____

29.

30.

29. _____

30. _____

31.

32.

33.

34.

35.

36.

37.

38.

39.

40.

31. _____

32. _____

33. _____

34. _____

35. _____

36. _____

37. _____

38. _____

39. _____

40. _____

Review Test #5 Lathes, Accessories, and Thread Cutting (Units 22 To 31)

91

41.

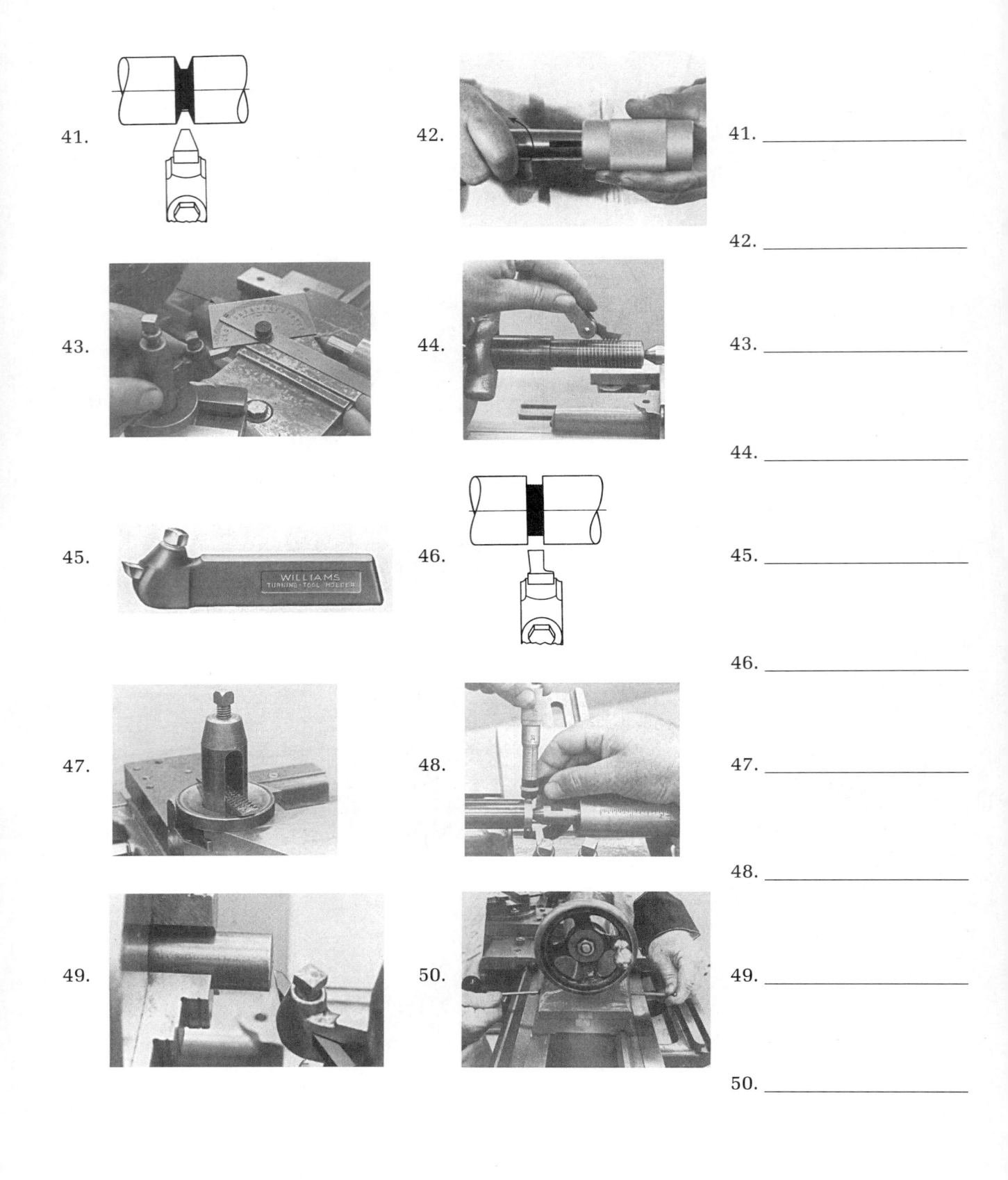

42.

43.

44.

45.

46.

47.

48.

49.

50.

41. _____

42. _____

43. _____

44. _____

45. _____

46. _____

47. _____

48. _____

49. _____

50. _____

Total—50 marks

# TEST #30
# HORIZONTAL MILLING MACHINES AND ACCESSORIES
## (SECTION 13, UNIT 32)

The milling machine is used to produce machined surfaces such as flats, angular surfaces, grooves, cams, contours, helical forms, and gear teeth. Milling operations are performed by feeding the stationary workpiece into a revolving cutter.

Place the correct word(s) in the space(s) at the right-hand side of the page to make the statement complete and true.

ANSWER

1. The type of cut a milling machine makes is determined by the size and __?__ of the __?__ .

    1. _____
       _____

2. The __?__ knee and column milling machine can perform a variety of operations by the addition of various __?__ .

    2. _____
       _____

3. The __?__ knee and column milling machine does not have a swivel table housing.

    3. _____

4. Special coatings can be applied to the surface of milling cutters to prolong tool __?__ , increase __?__ , and lower manufacturing __?__ .

    4. _____
       _____
       _____

5. The prime function of a coating is to reduce __?__ wear and produce a __?__ finish on the work surface.

    5. _____
       _____

6. Resistance to the chip flow causes small particles of metal to weld to the cutting tool face, forming a __?__ - __?__ __?__ .

    6. _____-_____
       _____

7. Titanium nitride coated tools last __?__ to __?__ times longer than noncoated tools.

    7. _____
       _____

8. __?__ milling is when the cutter rotation and the table feed are in the same direction.

    8. _____

9. __?__ milling is when the cutter rotation and the table feed are going in the opposite direction.

    9. _____

10. To reduce play and dimensional errors while machining, as much __?__ as possible should be removed from the table __?__ .

    10. _____
        _____

Test #30 Horizontal Milling Machines and Accessories (Section 13, Unit 32)

11. Be sure that the work and the cutter are mounted __?__ before taking a cut.

11. _____

12. Never attempt to mount, measure, or adjust work until the cutter is __?__ stopped.

12. _____

_____

Total—20 marks

# TEST #31
# MILLING MACHINE SETUP (SECTION 13, UNIT 33)

Because of the versatility of the milling machine and its attachments, it is possible to accurately machine parts of identical size and shape. For this to happen, the machine must be set up correctly and the work held securely.

Consider whether each statement is true or false, then indicate your choice by circling the proper letter in the right-hand column.

|     |     | ANSWER |
| --- | --- | --- |
| 1. | The milling arbor is held in the machine spindle by the draw-in bar. | 1. T F |
| 2. | The cutter is driven by the spacer and bearing bushings. | 2. T F |
| 3. | When removing an arbor, the end of the draw-in bar is hit with a hard-faced hammer until the arbor taper is free. | 3. T F |
| 4. | The teeth of a cutter should be pointing in the direction of arbor rotation when being mounted. | 4. T F |
| 5. | It is important that the arbor and arbor support clear the work. | 5. T F |
| 6. | Never tighten the arbor nut with a wrench unless the arbor support is in place. | 6. T F |
| 7. | Placing the cutter on the table surface will not damage the cutter accuracy. | 7. T F |
| 8. | Table vises are special work-holding devices made to hold one type of workpiece. | 8. T F |
| 9. | V-blocks usually have a 60° V-shaped groove to hold round work. | 9. T F |
| 10. | Angle plates are used for holding large work or squaring one surface with another. | 10. T F |
| 11. | Cutting speed is based on periphery speed expressed in either feet or meters per minute. | 11. T F |
| 12. | The cutting speeds used for milling machine cutters are the same as those used for any cutting tool. | 12. T F |
| 13. | Rigidity of the machine and workpiece does not affect the cutting speed used. | 13. T F |
| 14. | The formula for calculating cutting speed is the same for the inch and metric systems. | 14. T F |

15. Chip per tooth is the amount of material removed by each tooth of the cutter as it travels into the work.     15. T   F

16. Milling feed is calculated by multiplying the number of cutter teeth by the chip per tooth and the cutter r/min.     16. T   F

17. Twice as much material will be removed from the workpiece as is set on the graduated collars.     17. T   F

18. On milling machines, the value of graduated collar graduation is directly related to the diameter of the feed screw.     18. T   F

19. If the crossfeed screw on the saddle has five threads per inch, one complete revolution of the screw advances the saddle .020 in.     19. T   F

20. A crossfeed screw with a pitch of 5 mm advances the saddle 5 mm in one revolution.     20. T   F

Total—20 marks

# TEST #32
# HORIZONTAL MILLING OPERATIONS (SECTION 13, UNIT 34)

Due to the number of attachments and cutting tools available, the horizontal milling machine is a very versatile and flexible machine tool, capable of performing a wide variety of machining operations.

Select the correct answer for each question and indicate your choice by circling the correct letter in the right-hand column.

ANSWER

1. In order to machine square or parallel to an edge, the vise jaws must first be
    (A) changed
    (B) aligned
    (C) altered
    (D) reversed

    1.  A   B   C   D

2. To prevent accidents and damage to the machine and workpiece, the work must be accurately set up and held
    (A) snugly
    (B) lightly
    (C) securely
    (D) weakly

    2.  A   B   C   D

3. The most accurate method of aligning a vise is by using
    (A) matching lines
    (B) a steel square
    (C) an indicator
    (D) a test bar

    3.  A   B   C   D

4. The safest way to set the cutter to the work surface is by using
    (A) long paper strip
    (B) eye judgment
    (C) micrometer
    (D) feelers

    4.  A   B   C   D

5. To ensure that work held in a vise is properly seated on parallels, all corners of the work are tapped down on
    (A) parallels
    (B) paper feelers
    (C) matching lines
    (D) none of these

    5.  A   B   C   D

6. Which of the following cutters is recommended to mill a flat surface?
    (A) staggered-tooth
    (B) plain
    (C) side
    (D) helical

    6.  A   B   C   D

7. Once the cutter has made contact with the workpiece, depth of cut should be set by the

(A) steel rule
(B) indicator
(C) graduated collar
(D) micrometer

7. A B C D

8. For roughing cuts, the depth of cut should not be less than

(A) 1/32 in. (0.75 mm)
(B) 1/16 in. (1.5 mm)
(C) 3/32 in. (2.25 mm)
(D) 1/8 in. (3 mm)

8. A B C D

9. The depth of the finish cut should be between

(A) .005 to .008 in. (0.12 to 0.2 mm)
(B) .010 to .025 in. (0.25 to 0.65 mm)
(C) .015 to .035 in. (0.38 to 0.89 mm)
(D) .025 to .050 in. (0.64 to 1.27 mm)

9. A B C D

10. To prevent damage to the cutter, vise, or parallels, the work should extend beyond the edge of the vise by about

(A) 1/8 in. (3 mm)
(B) 1/4 in. (6 mm)
(C) 3/8 in. (9 mm)
(D) 1/2 in. (12 mm)

10. A B C D

Consider whether each statement is true or false, then indicate your choice by circling the proper letter in the right-hand column.

11. Keyways and slots may be cut using a side milling or slotting cutter. — 11. T F
12. The dividing head is another term for index head. — 12. T F
13. The spindle rotation on the index head is related to a 50-tooth worm wheel. — 13. T F
14. Work may be indexed by either simple or indirect indexing. — 14. T F
15. The index plate and sector arms are used together to divide work circumference into fractional parts. — 15. T F
16. When counting holes for indexing, the hole in which the index crank pin is engaged should not be included. — 16. T F

17. In setting the hole location in the index plate, the index crank should be turned counterclockwise.   17.  T   F

18. When the index pin has been moved past the required hole, the backlash must be removed to prevent an error in spacing.   18.  T   F

19. Direct indexing can be used for milling 22, 28, or 34 divisions.   19.  T   F

20. Before setting any depth of cut, the work should be moved clear of the cutter.   20.  T   F

Total—20 marks

# TEST #33
# VERTICAL MILLING MACHINES (SECTION 13, UNIT 35)

The vertical milling machine is basically the same as the horizontal plain milling machine, but has a vertical spindle. It can perform operations such as face and end milling, cutting keyways, dovetails, T-slots, and also drilling, boring, and jig boring.

Place the correct word(s) in the space(s) at the right-hand side of the page to make the statement complete and true.

ANSWER

1. A __?__ end mill is used for milling a __?__ radius on the bottom of slots.  
   1. _____ _____

2. The corner-rounding end mill is used to produce a __?__ form on the corner of a workpiece.  
   2. _____

3. The __?__ cutter is used for milling dovetail slides into a workpiece.  
   3. _____

4. Four-flute end mills are not __?__ -cutting and are generally used for __?__ cutting.  
   4. _____ _____

5. __?__ - __?__ end mills minimize chatter and provide good chip clearance.  
   5. _____ - _____

6. The __?__ provides an economic way to machine large work surfaces.  
   6. _____

7. There are two main types of collets, the __?__ type and the __?__ type.  
   7. _____ _____

8. Excessive __?__ is one of the main causes of cutting edge __?__ and shortened tool life.  
   8. _____ _____

9. __?__ is a wearing-away action caused by the __?__ of the material being cut.  
   9. _____ _____

10. __?__ occurs on cutting edges when the __?__ cutting forces are greater than the cutting edges can stand.  
    10. _____ _____

11. A cutter should never be allowed to __?__ on the work to prevent work __?__.  
    11. _____ _____

12. Cratering is the creation of a narrow __?__ in the tooth face which eventually results in tooth failure.  
    12. _____

13. __?__ - __?__ end mills are used for general purpose milling.  
    13. _____ - _____

Total—20 marks

# TEST #34
# VERTICAL MILLING MACHINE OPERATIONS
# (SECTION 13, UNIT 36)

The vertical milling machine can perform many types of operations because of the wide variety of cutting tools and attachments that are available. It is especially useful for drilling, boring, reaming, and jig boring operations.

Consider whether each statement is true or false, then indicate your choice by circling the proper letter in the right-hand column.

ANSWER

1. Tools inserted in a collet or adapter are held in the spindle by means of a draw-in bar.    1. T F

2. The vertical head should be at 90° to the table in two directions for machining a flat surface.    2. T F

3. Since the work will be machined, burrs do not have to be removed when mounting the work.    3. T F

4. When a block is machined square and parallel, the sides can be cut in any order.    4. T F

5. The ends of a workpiece may be machined by any of three methods.    5. T F

6. When end machining short pieces not more than 3 1/2 in. (90 mm) long, they may be held upright in the middle of a vise.    6. T F

7. The vise jaw must be set parallel to the table travel for milling the end of a workpiece.    7. T F

8. It is important to use paper feelers between the work and parallels when setting work in a vise.    8. T F

9. When machining an end, the workpiece should extend at least 1 in. (25 mm) beyond the vise jaws.    9. T F

10. A paper feeler should be used when setting the rotating cutter to the workpiece.    10. T F

Place the correct word(s) in the space(s) at the right-hand side of the page to make the statement complete and true.

11. When the vertical head is at right angles (90°) to the table, the head and table are considered to be __?__ .    11. _____

12. The vise jaw should be __?__ parallel to the table __?__ to produce accurate work.

13. The round bar must be in the __?__ of the work held __?__ the vise to hold it squarely against the solid jaw.

14. When drilling, work should be set on __?__ which will not interfere with the __?__ .

15. After the hole has been drilled, the table position should not be __?__ if the hole is to be reamed.

16. The speed for reaming is approximately __?__ of the drilling speed.

17. The reamer should never be turned __?__ when removing it from the hole.

12. _____
    _____
13. _____
    _____
14. _____
    _____
15. _____
16. _____
17. _____

_____

Total—20 marks

# REVIEW TEST #6
# MILLING MACHINES (SECTION 13—UNITS 32 TO 36)

**PART 1**

Place the name of each milling machine, accessory, cutting tool, or operation in the proper space in the right-hand column.

1. spring collet  _____

2. conventional milling  _____

3. dovetail cutter  _____

4. half-side milling  _____

5. plain milling  _____

6. climb milling  _____

7. vertical milling machine  _____

8. flycutter  _____

9. shell end mill  _____

10. gear cutter  _____

11. horizontal milling machine _____

12. setting cutter to work  _____

13. heavy duty plain  _____

14. convex  _____

Review Test #6 Milling Machines (Section 13—Units 32 to 36)

105

15. deburring _____

16. corner rounding _____

17. single angle _____

18. roughing _____

19. two-flute _____

20. four-flute _____

21. T-slot _____

22. ball _____

23. side milling _____

24. solid collet _____

25. removing chips _____

26. concave _____

106     Lab Manual to Accompany Machine Tool and Manufacturing Technology

## PART 2

Select the correct answer for each question and indicate your choice by circling the correct letter in the right-hand column.

ANSWER

27. Coated cutting tools can increase tool life as much as 27.  A  B  C  D
    (A) 100%   (C) 300%
    (B) 200%   (D) 400%

28. Coated cutting tools can result in increased productivity by as much as 28.  A  B  C  D
    (A) 75%    (C) 60%
    (B) 50%    (D) 90%

29. The milling operation for which half-side milling cutters are most commonly used is 29.  A  B  C  D
    (A) straddle   (C) plain
    (B) concave    (D) heavy duty

30. The inch r/min formula for milling cutters is 30.  A  B  C  D
    (A) $\dfrac{CS \times 320}{D}$   (C) $\dfrac{CS \times 4}{D}$
    (B) $\dfrac{N \times c.p.t.}{feed}$   (D) $\dfrac{N \times c.p.t.}{mm/min}$

31. The metric r/min formula for milling cutters is 31.  A  B  C  D
    (A) $\dfrac{CS \times 320}{D}$   (C) $\dfrac{CS \times 4}{D}$
    (B) $\dfrac{CS \times D}{320}$   (D) $\dfrac{D \times 320}{CS}$

Review Test #6 Milling Machines (Section 13—Units 32 to 36)

## PART 3

Place the correct word(s) in the space(s) at the right-hand side of the page to make the statement complete and true.

ANSWER

32. The simplest, but least accurate method of aligning a vise is to match the __?__ on the vise and the __?__ base.

32. _____
    _____

33. The most accurate method of aligning a vise is with an __?__ .

33. _____

34. To remove backlash, turn the handle __?__ one-half turn and then come to the graduation line.

34. _____

35. __?__ is the rate at which the work moves lengthwise into the revolving __?__ .

35. _____
    _____

36. The head of a vertical mill can be aligned using a __?__ on the table and checking the spindle sleeve.

36. _____

37. To ensure that work seats firmly on parallels, __?__ __?__ are placed between the parallels and the work.

37. _____
    _____

38. __?__ collars are placed on the end of __?__ screws to accurately measure movement.

38. _____
    _____

39. The maximum __?__ of cut for a vertical milling cutter should be no more than the __?__ of the end mill.

39. _____
    _____

40. Jamming the flutes of an end mill with gummy material is known as __?__ .

40. _____

41. To prevent a built-up edge from forming, reduce the table __?__ , depth of cut, and apply __?__ fluid.

41. _____
    _____

42. Cratering is caused by the high __?__ and abrasion of the __?__ sliding on the __?__ face next to the cutting edge.

42. _____
    _____
    _____

_____

Total—50 marks

# TEST #35
# BENCH AND ABRASIVE BELT GRINDERS (SECTION 14, UNIT 37)

Grinding is a metal removal process which uses abrasive grains that are found on the periphery of a rotating grinding wheel or an abrasive belt. These grains act as cutting tools which are capable of finishing soft or hardened work surfaces to extremely close tolerances.

Consider whether each statement is true or false, then indicate your choice by circling the proper letter in the right-hand column.

ANSWER

1. The bench or pedestal grinder is used for sharpening tools. 1. T F
2. Abrasive belt grinders can be used for finishing flat or contour work. 2. T F
3. Offhand grinding is another term for belt grinding. 3. T F
4. The two wheels found on a pedestal grinder are usually coarse-grained wheels. 4. T F
5. Aluminum oxide and silicon carbide are not as hard as natural abrasives. 5. T F
6. Natural abrasives such as quartz, emery, and sandstone contain no impurities. 6. T F
7. Silicon carbide is manufactured in an arc-type electric furnace. 7. T F
8. A grinding wheel should be ring tested before mounting. 8. T F
9. The work rest should be adjusted within 1/8 in. (3 mm) of the wheel. 9. T F
10. The wheel guard should cover at least one-third of the grinding wheel. 10. T F
11. When the grinder is equipped with eyeshields, it is not necessary to wear safety glasses. 11. T F
12. Never stand in line with a grinding wheel during the start-up. 12. T F
13. If a wheel is to shatter during start-up, it will break within the first minute. 13. T F
14. Dressing or truing a wheel both mean the same thing. 14. T F
15. Wheel loading refers to small metal particles that imbed themselves in the wheel. 15. T F

16. A mechanical dresser is generally used to dress off-hand grinding wheels.  16. T  F
17. In offhand grinding, the toolbit is held by mechanical means.  17. T  F
18. All lathe toolbits must have relief and rake angles for them to cut effectively.  18. T  F
19. Once a general-purpose lathe toolbit has been ground, it can be quickly resharpened by grinding the top rake.  19. T  F
20. A line marked at 49° on the grinder work rest will keep a standard drill at the correct angle.  20. T  F

_____

Total—20 marks

# TEST #36
# SURFACE GRINDER WHEELS AND OPERATIONS
# (SECTION 14, UNIT 38)

The surface grinder, used primarily for grinding flat surfaces on hardened or unhardened workpieces, has become a very valuable machine tool in industry because it can also grind contours and shapes to high finishes and close tolerances.

Place the correct word(s) in the space(s) at the right-hand side of the page to make the statement complete and true.

ANSWER

1. The most common surface grinder is the __?__ spindle grinder with a __?__ table.

1. _____
   _____

2. A __?__ chuck provides a fast and easy method of __?__ the work while grinding.

2. _____
   _____

3. The wheelfeed handwheel moves the grinding wheel __?__ or __?__ to set the depth of cut.

3. _____
   _____

4. A wheel that is run faster than recommended can __?__ and cause an accident or __?__ the machine.

4. _____
   _____

5. Never attempt to __?__ or __?__ work until the grinding wheel has completely stopped.

5. _____
   _____

6. The grade of a grinding wheel generally refers to the __?__ of the __?__ which hold the abrasive grains together.

6. _____
   _____

7. The structure of a grinding wheel depends on the __?__ taken up by the __?__ and the __?__ compared to the voids between them.

7. _____
   _____
   _____

8. A wheel should always be inspected when received to check for __?__ and __?__ and then __?__ -tested.

8. _____
   _____
   _____

9. Thin wheels should be stored on a flat horizontal surface to prevent __?__ .

9. _____

10. Excessive pressure should be avoided when tightening the wheel because __?__ may be created that can cause the wheel to __?__ .

10. _____
    _____

11. __?__ is the operation of removing dull grains to make the wheel cut better.

11. _____

Test #36 Surface Grinder Wheels and Operations (Section 14, Unit 38)

12. The operation that changes the face to give it a desired shape is called __?__ .  12. _____

13. To prevent scratching or marring the chuck, __?__ should be placed between the workpiece and the chuck.  13. _____

14. A clean-cutting wheel should produce a __?__ surface finish on the work.  14. _____

_____

Total—25 marks

# REVIEW TEST #7
# GRINDING OPERATIONS (SECTION 14, UNITS 37 AND 38)

**PART I**

Consider whether each statement is true or false, then indicate your choice by circling the proper letter in the right-hand column.

|     |                                                                                                                                  |     | ANSWER |     |
| --- | -------------------------------------------------------------------------------------------------------------------------------- | --- | ------ | --- |
| 1.  | A fine-grained grinding wheel is used for the rapid removal of metal.                                                            | 1.  | T      | F   |
| 2.  | Manufactured abrasives generally don't perform as well as natural abrasives, such as emery, sandstone, corundum, and quartz.     | 2.  | T      | F   |
| 3.  | A loaded wheel cuts properly, but the process takes much longer than it would with a properly dressed wheel.                     | 3.  | T      | F   |
| 4.  | Dressing is the process of reconditioning a wheel to improve its grinding performance.                                           | 4.  | T      | F   |
| 5.  | Aluminum oxide abrasive belts should be used to grind materials that have high tensile strength.                                 | 5.  | T      | F   |
| 6.  | It's not necessary to wear safety goggles when grinding soft materials.                                                          | 6.  | T      | F   |
| 7.  | Grinding wheels are made up of two materials, the bond or base material and the cement that holds it together.                   | 7.  | T      | F   |
| 8.  | The hardest wheel grade is "A" and the softest grade is "Z."                                                                     | 8.  | T      | F   |
| 9.  | When ring testing a grinding wheel, a strong dull sound indicates that the wheel is in good condition.                           | 9.  | T      | F   |
| 10. | The most common operation performed on a surface grinder is grinding flat or horizontal surfaces.                                | 10. | T      | F   |

## PART 2

Place the correct word(s) in the space(s) at the right-hand side of the page to make the statement complete and true.

ANSWER

11. Grinding wheels operate at very high __?__ and the grinding particles are very __?__ .

11. _____
_____

12. You should allow a new grinding wheel to run for about __?__ minute(s) before using it.

12. _____

13. Adjustments should not be made on a grinder unless the wheel is __?__ .

13. _____

14. High-speed toolbits must be __?__ frequently.

14. _____

15. Before starting a grinder, be sure that the wheel __?__ the work.

15. _____

16. Diamond abrasives should never be used on __?__ metals.

16. _____

17. The __?__ of the abrasive grain affects the type of surface finish that is produced.

17. _____

18. Two factors to consider in selecting a grinding wheel grade are the wheel speed and the __?__ rate.

18. _____

19. File off any __?__ on the surface of work that is to be placed on the magnetic chuck.

19. _____

_____

Total—20 marks

# TEST #37
# COMPUTER NUMERICAL CONTROL (SECTION 15, UNIT 39)

Computers are being used to operate and control machines at the maximum speeds possible, while at the same time being able to check the accuracy of the product and take corrective action, if necessary. No other invention in history has had such an impact on humanity in such a short period of time.

Place the correct word(s) in the space(s) at the right-hand side of the page to make the statement complete and true.

ANSWER

1. The __?__ was really the first primitive computer and it was developed in Asia.

    1. _____

2. The __?__ card system, the first method of data processing, was developed in the early 1800s.

    2. _____

3. The computer is a tool which can perform many tasks with amazing speed, __?__ , and __?__ .

    3. _____
    _____

4. The computer is a machine consisting of an arithmetic __?__ unit, a __?__ unit, and an input/__?__ device that can process data.

    4. _____
    _____
    _____

5. The three most common types of computers are __?__ , __?__ , and __?__ .

    5. _____
    _____
    _____

6. A computer system usually has both __?__ and __?__ memory storage.

    6. _____
    _____

7. Some of the more common CNC machine tools are turning and __?__ centers, __?__ and vertical machining centers, coordinate __?__ machines, and __?__-__?__ machines.

    7. _____
    _____
    _____
    _____-_____

8. Turning centers operate primarily on the __?__ and __?__ axes.

    8. _____
    _____

9. Vertical machining centers operate on the __?__ , __?__ , and __?__ axes.

    9. _____
    _____
    _____

Test #37 Computer Numerical Control (Section 15, Unit 39)

10. The wide acceptance of CNC machine tools is due to their __?__ , __?__ , __?__ and productivity.

19. _____
    _____
    _____

    _____
    Total—25 marks

# TEST #38
# HOW CNC CONTROLS MACHINES (SECTION 15, UNIT 40)

A variety of input media can be used to control the movement of CNC machine tool slides to perform the functions necessary to machine parts. These slides can move:
- independently on straight line axes which are at right angles to each other.
- in angular motions where two slides move at the same time.
- in circular motions where generally two slides make many thousands of minute (very small) moves to produce curved forms.

Place the correct word(s) or information in the space(s) at the right-hand side of the page to make the statement complete and true.

ANSWER

1. The Cartesian coordinate system works on a __?__ system, where reference __?__ run at right angles to each other.

   1. _____
   _____

2. An X __?__ dimension would be to the right of the Y coordinate line.

   2. _____

3. A Y __?__ dimension would be below the X coordinate line.

   3. _____

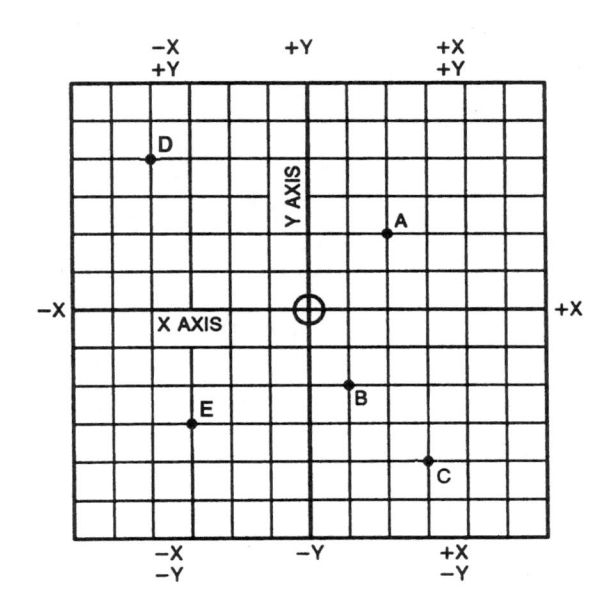

4. Give the XY coordinate locations for each point on the diagram.

   4A. _____

   B. _____
   C. _____
   D. _____
   E. _____

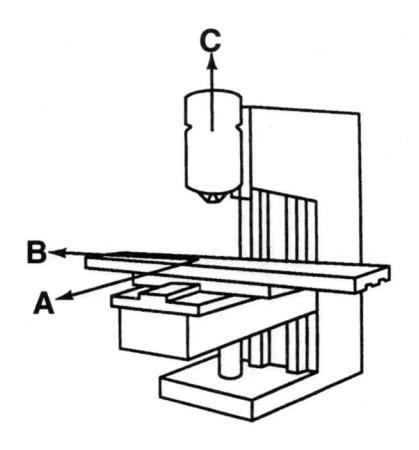

5. Label the vertical milling machine axes indicated by the arrows.

5A. _____

B. _____

C. _____

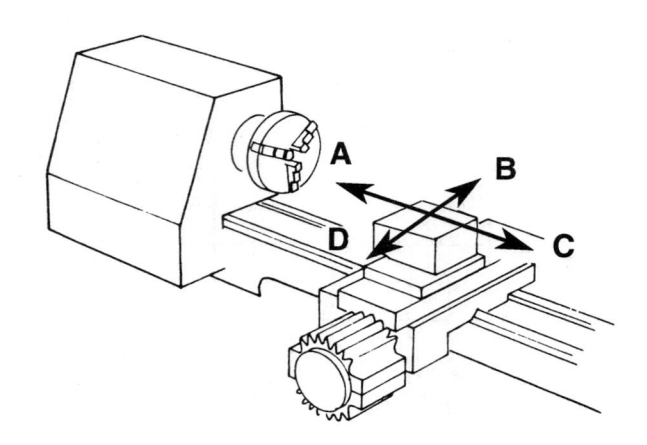

6. Label the lathe axes indicated by the arrows.

6A. _____

B. _____

C. _____

D. _____

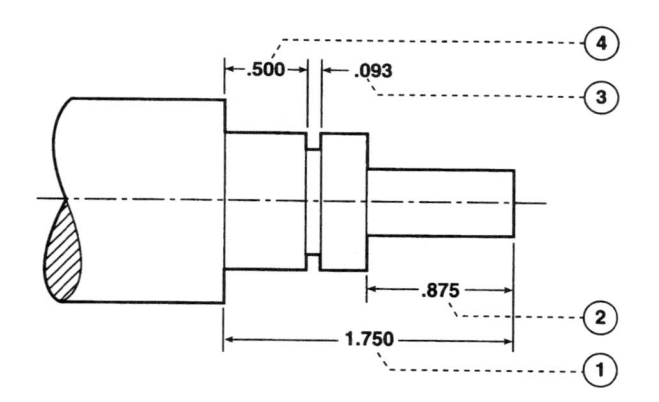

7. What programming mode is being used for:

   (A) Points 1 and 2

   (B) Points 3 and 4

   7A. _____

   B. _____

8. Continuous-path machining or __?__ , is where the cutting tool is __?__ in contact with the workpiece as it travels from one programmed point to the next.

   8. _____
   _____

9. The __?__ address is the most common type of programming format used for CNC programming.

   9. _____

10. The most common codes used when programming CNC machine tools are __?__ codes and __?__ codes.

    10. _____
    _____

11. __?__ codes stay in effect in a program until they are changed by another code from the same __?__ number.

    11. _____
    _____

Total—25 marks

Test #38 How CNC Controls Machines (Section 15, Unit 40)

# TEST #39
# PREPARING FOR PROGRAMMING (SECTION 15, UNIT 41)

Accurate programming is very important to how accurate the CNC machine produces a part. Regardless of the method of programming, it is very important for a programmer to have a good knowledge of print reading, machining sequences, cutting tools, and work-holding devices. Before any programming can begin, a reference point must be established for the workpiece and the machine tool.

Place the correct word(s) of information in the space(s) at the right-hand side of the page to make the statement complete and true.

ANSWER

1. The programmer must be able to plan the correct __?__ sequences for a part from the technical or engineering drawing.

   1. _____

2. Identify the type of each command on the programmed line.

   N6 G01 X6.0 Z-1.250 S1000 F5

   2. N _____
   G _____
   X _____
   Z _____
   S _____
   F _____

3. The G20 code is used for __?__ dimension programming; the G21 code is used for __?__ programming.

   3. _____
   _____

4. Point-to-point or __?__ is used to move a tool __?__ to a specific location on a workpiece where some operation will be performed.

   4. _____
   _____

5. __?__ interpolation is used for straight-line moves.

   5. _____

6. A continuous path or __?__ operation is where the cutting tool is __?__ in contact with the part as curved movements are made.

   6. _____
   _____

7. Machine and work coordinates are used as reference points when locating __?__ , __?__ , devices, and for __?__ - __?__ positions.

   7. _____
   _____
   _____

8. Offsets or compensation are used to make adjustments in the program for tool __?__ , __?__ , nose radius, __?__ , and deflection.

8. _____
   _____
   _____

9. In cutter radius compensation, the G __?__ code is used when the cutter is to the left of the part, the G __?__ when the cutter is to the right, and G __?__ to cancel cutter radius compensation.

9. _____
   _____
   _____

10. All __?__ coordinate dimensions are taken from one fixed point.

10. _____

11. Each __?__ coordinate dimension is taken from the previous point.

11. _____

_____

Total—25 marks

# TEST #40
# LINEAR PROGRAMMING (SECTION 15, UNIT 42)

Linear interpolation involves the straight-line movement of machine tool slides between any two programmed points, whether they are close together or far apart. Straight-line angular cuts can be made with the G01 command when the X and Y coordinates at the beginning and end of the line are given along with a programmed feed rate. The G00 command is used for rapid positioning between two points, while the G01 command is used for straight-line cuts.

For the following programming exercises, supply the missing information for each program in the spaces in the right-hand column.

1. **Programming the PART BOUNDARY and the location of the THREE HOLES**
   *Program Notes*
   - Only program the part boundary and hole locations in the rapid traverse (G00) mode.
   - Locate the spindle at the start point (tool-change position) for all changes.
   - All programming for the part begins at the XY zero (lower left corner of the part).
   - Use incremental programming for:
     (a) the part boundary in a clockwise direction.
     (b) the hole locations in numerical order.
   - After the hole #3 position, return to the start point.

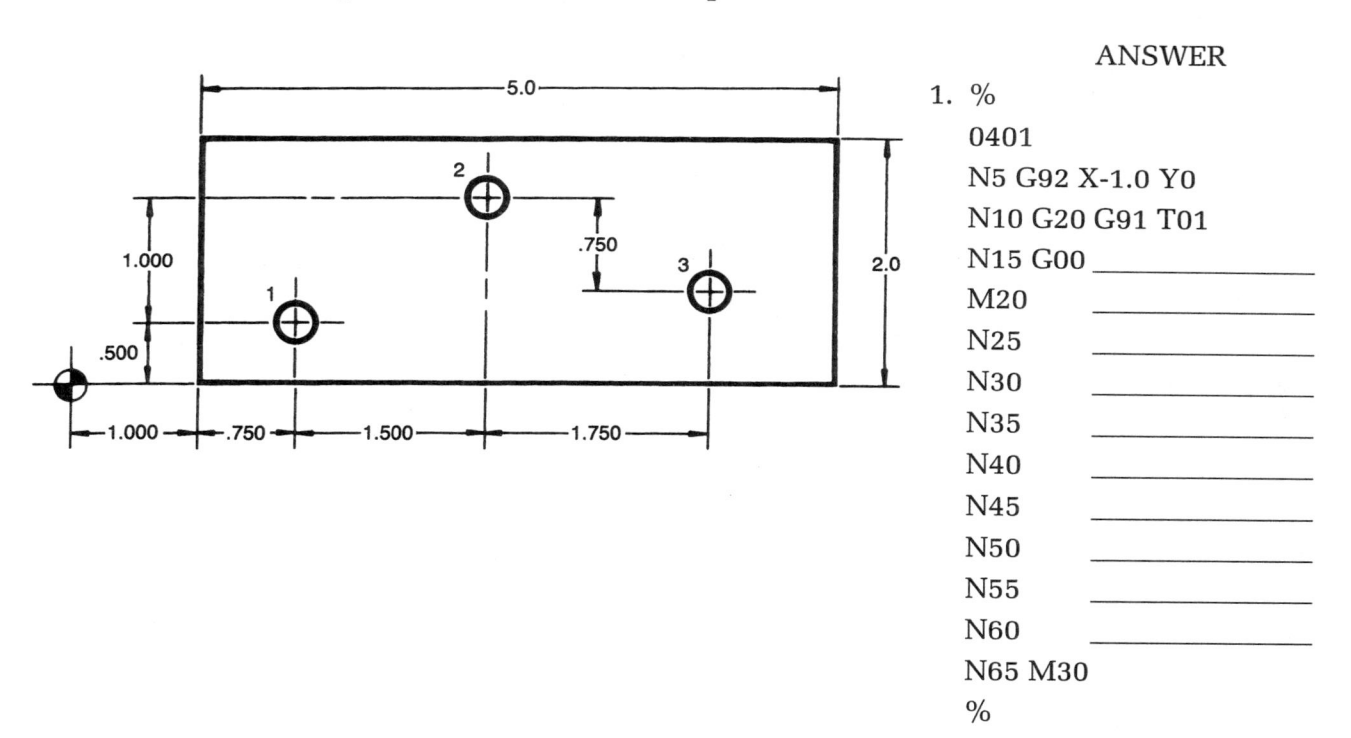

ANSWER

1. %
   0401
   N5 G92 X-1.0 Y0
   N10 G20 G91 T01
   N15 G00 _____
   M20 _____
   N25 _____
   N30 _____
   N35 _____
   N40 _____
   N45 _____
   N50 _____
   N55 _____
   N60 _____
   N65 M30
   %

2. **Programming the location of the FIVE HOLES**
   *Program Notes*
   - Only program the location of all five holes in numerical sequence in the rapid traverse mode.
   - Locate the spindle at the start point (tool-change position).
   - All programming for the part begins at the XY zero (top right corner of the part).
   - Use absolute positioning to locate the spindle at each hole in numerical order.
   - After hole #5, return to the start point.

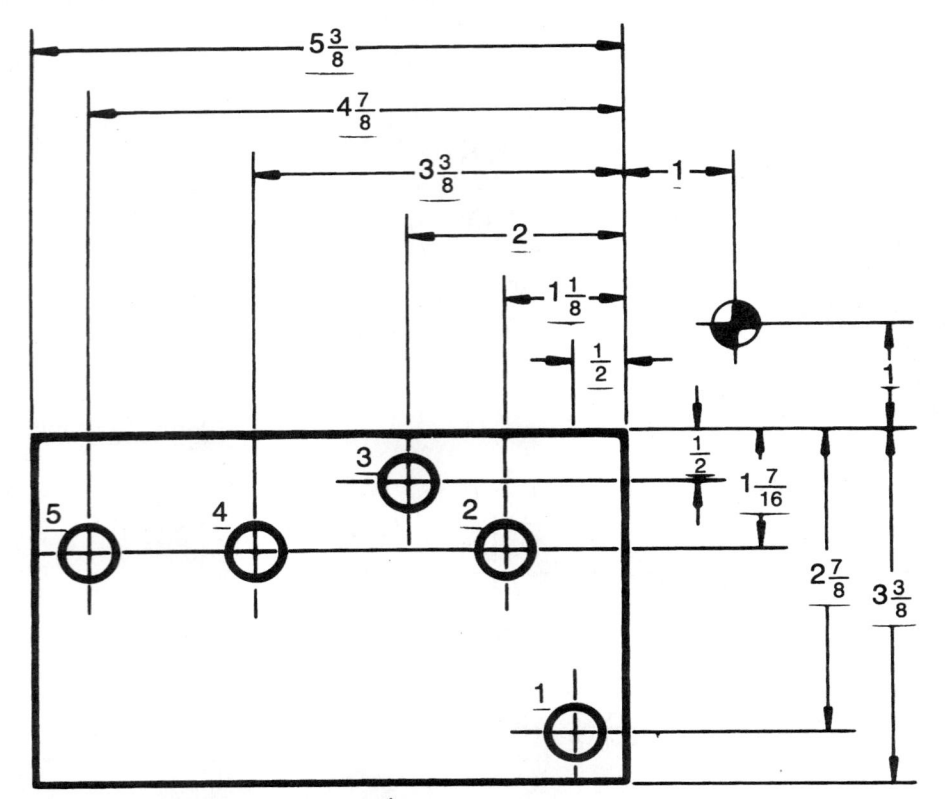

DRILL 5 HOLES THRU – $\frac{1}{2}$ DIA.

ANSWERS

2. %
   0402
   N5 G92 X1.0 Y1.0
   N10_____
   N15_____
   N20_____
   N25_____
   N30_____
   N35_____
   N40_____
   N45_____
   N50_____
   %

3. **Programming the ANGULAR SLOT**

   *Program Notes*
   - Program in the absolute mode to mill only the slot.
   - Locate the spindle at the start position.
   - All programming for the part starts at the XY zero (bottom left corner of part).
   - Start the slot, 1/4 in. wide and 1/4 in. deep, at position A.
   - At the completion of the slot, return to the start point.
   - The material is aluminum (CS 300). Feed is 5 in./min.

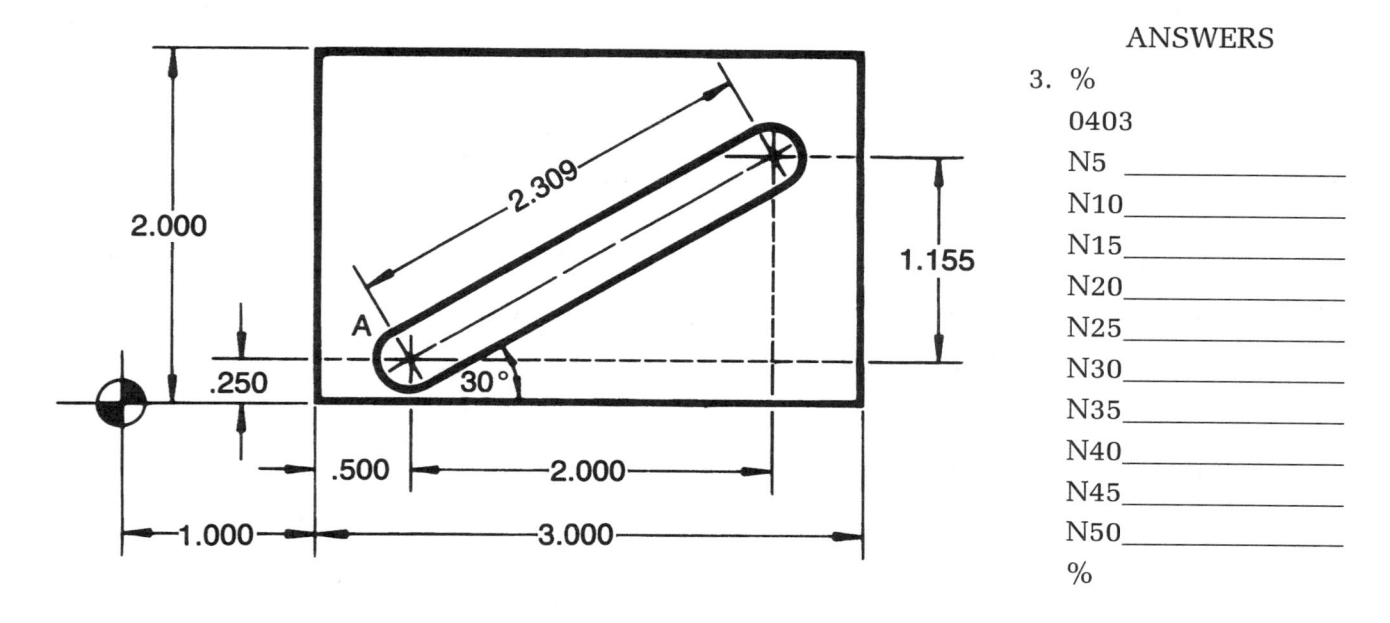

   ANSWERS
   3. %
   0403
   N5 _____
   N10 _____
   N15 _____
   N20 _____
   N25 _____
   N30 _____
   N35 _____
   N40 _____
   N45 _____
   N50 _____
   %

## 4. Milling and Drilling Project

Fill in the missing information, indicated by the numbers in brackets, in the correct spaces provided in the right-hand column.

*Program Notes*
- Program in the absolute mode starting at the XY zero (top left corner of the part).
- Material is aluminum (300 CS), feed rate 10 in., high speed steel tools.
- Drill the two 1/4 in. diameter holes through the workpiece.
- Return to the start position for the tool change.
- Mill the 1/4 in. wide square slot clockwise starting at point A.
- Return to the start point.

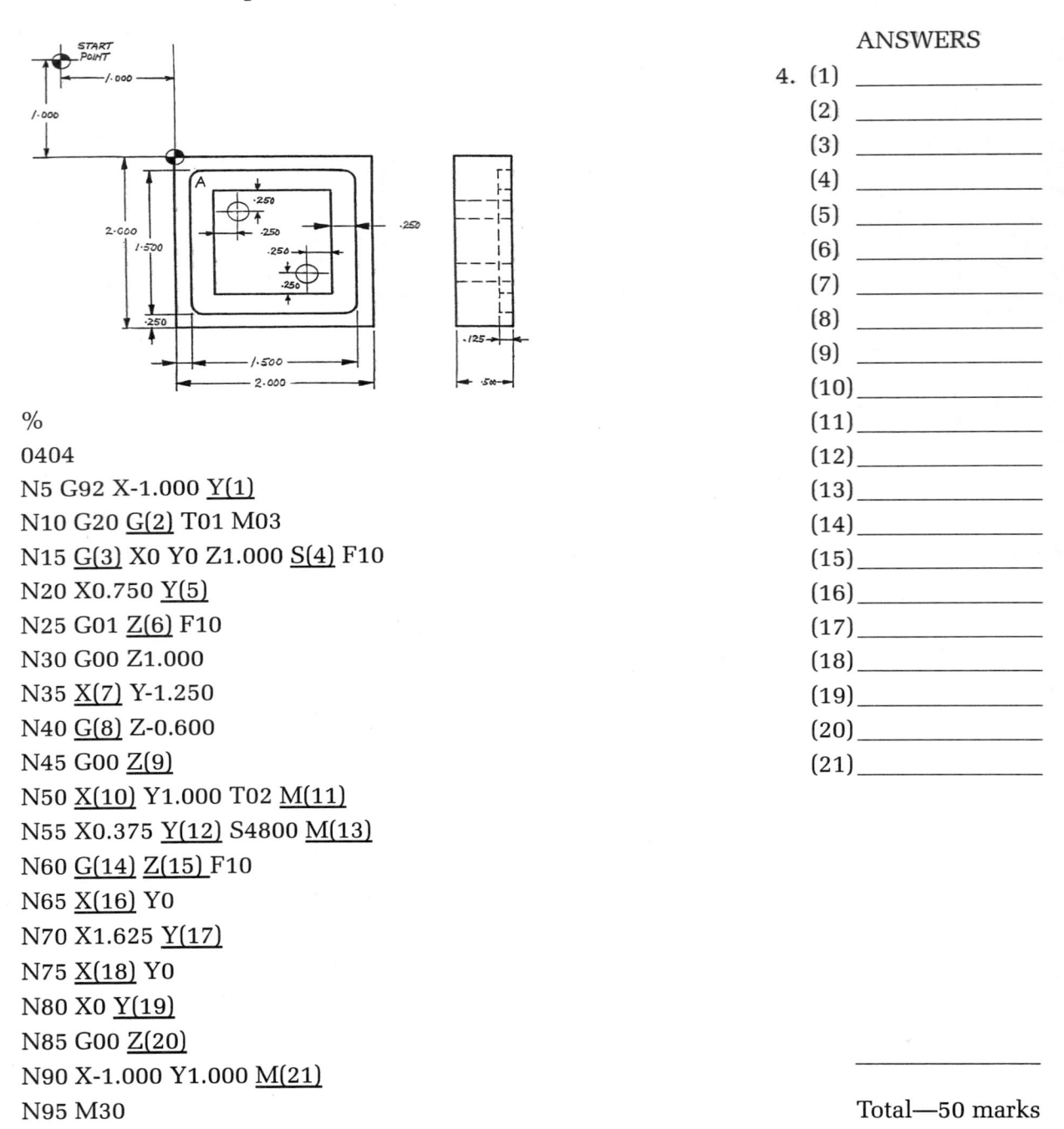

ANSWERS

4. (1) _____
(2) _____
(3) _____
(4) _____
(5) _____
(6) _____
(7) _____
(8) _____
(9) _____
(10) _____
(11) _____
(12) _____
(13) _____
(14) _____
(15) _____
(16) _____
(17) _____
(18) _____
(19) _____
(20) _____
(21) _____

```
%
0404
N5 G92 X-1.000 Y(1)
N10 G20 G(2) T01 M03
N15 G(3) X0 Y0 Z1.000 S(4) F10
N20 X0.750 Y(5)
N25 G01 Z(6) F10
N30 G00 Z1.000
N35 X(7) Y-1.250
N40 G(8) Z-0.600
N45 G00 Z(9)
N50 X(10) Y1.000 T02 M(11)
N55 X0.375 Y(12) S4800 M(13)
N60 G(14) Z(15) F10
N65 X(16) Y0
N70 X1.625 Y(17)
N75 X(18) Y0
N80 X0 Y(19)
N85 G00 Z(20)
N90 X-1.000 Y1.000 M(21)
N95 M30
%
```

Total—50 marks

# TEST #41
# CIRCULAR INTERPOLATION (SECTION 15, UNIT 43)

Circular interpolation was developed to eliminate the many calculations required with early controls to program arcs and circles. The circular interpolation unit of the MCU automatically breaks up an arc into very small (minute) linear moves to produce a circular path. Some machine control units generate only one arc at a time and therefore it takes four programmed lines (commands) to produce a circle. Modern controls can generate arcs and circles by the radius method when the arc radius and the XY coordinate dimensions are provided.

Place the correct word(s) or information in the space(s) at the right-hand side of the page to make the statement complete and true.

ANSWERS

1. The most common types of interpolation are __?__ , circular, and __?__ .

1. _____
   _____

2. The information required to program an arc or circle on some controls are the XY coordinates for the __?__ and __?__ point of the arc, the __?__ of cutter travel, and the __?__ point of the arc.

2. _____
   _____
   _____
   _____

3. The code for circular interpolation clockwise is __?__ ; for counterclockwise the __?__ is used.

3. _____
   _____

4. The three factors which are required for radius programming are __?__ code, __?__ point of the arc, and the __?__ of the circle.

4. _____
   _____
   _____

Test #41 Circular Interpolation (Section 15, Unit 43)

## 5. Machining a 90° Arc (Center-Point Method)

Fill in the missing information, indicated by the numbers in brackets, in the correct spaces in the right-hand column.

*Program Notes*
- Program in the absolute mode starting at the XY zero (lower left corner of the part).
- Machine the path from Point #1 to 4, .250 in. deep; material is machine steel (CS 100).
- High speed steel 1/2 in. diameter end mill.
- Return to tool-change position.

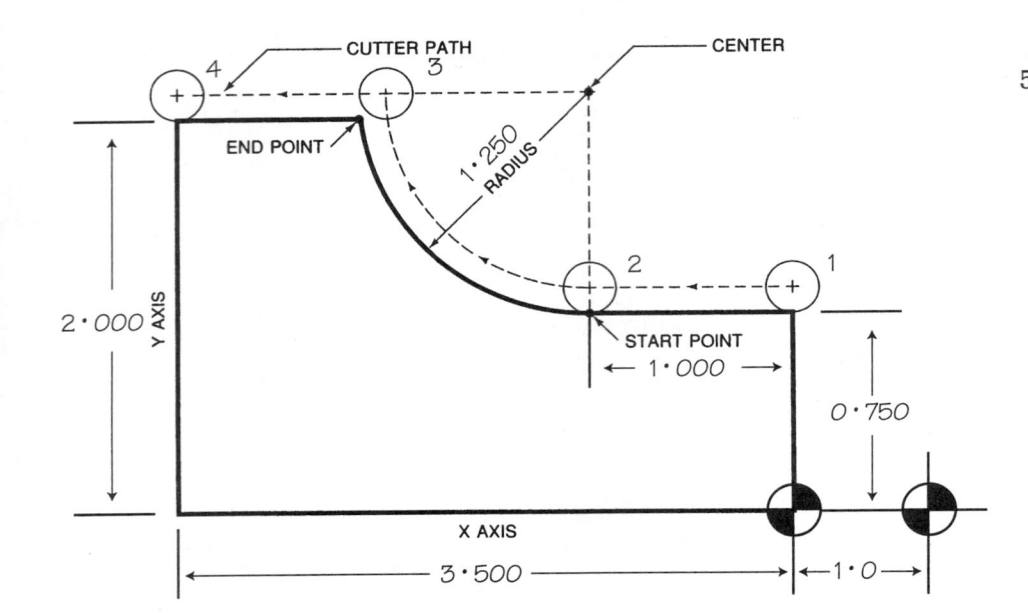

ANSWERS

5. (1) _____
(2) _____
(3) _____
(4) _____
(5) _____
(6) _____
(7) _____
(8) _____
(9) _____
(10) _____
(11) _____
(12) _____
(13) _____
(14) _____
(15) _____
(16) _____
(17) _____

```
%
04371
N5 G54 G17 G(1) G40 G(2)
N10 T01 M(3)
N15 G43 H01 Z(4)
N20 S(5) M03
N25 G00 X.250 Y(6) Z.1
N30 G(7) Z(8) F5.0
N35 X(9)
N40 G(10) X-2.000 Y(11) I0 J(12)
N45 G01 X(13)
N50 G00 Z(14) G49 M(15)
N55 X(16) Y0
H60 M(17)
%
```

## ANSWER 6. Machining a FULL CIRCLE (Center-Point Method)

Fill in the missing information, indicated by the numbers in brackets, in the correct spaces in the right-hand column.

*Program Notes*
- Use absolute programming and begin at the XY zero (start point to the left).
- Use a 1/4 in., two-flute end mill to cut the circular groove clockwise .125 in. deep, programming each quadrant individually.
- The material is aluminum (CS 300).
- After machining the full circle (point #5), return to the start point.

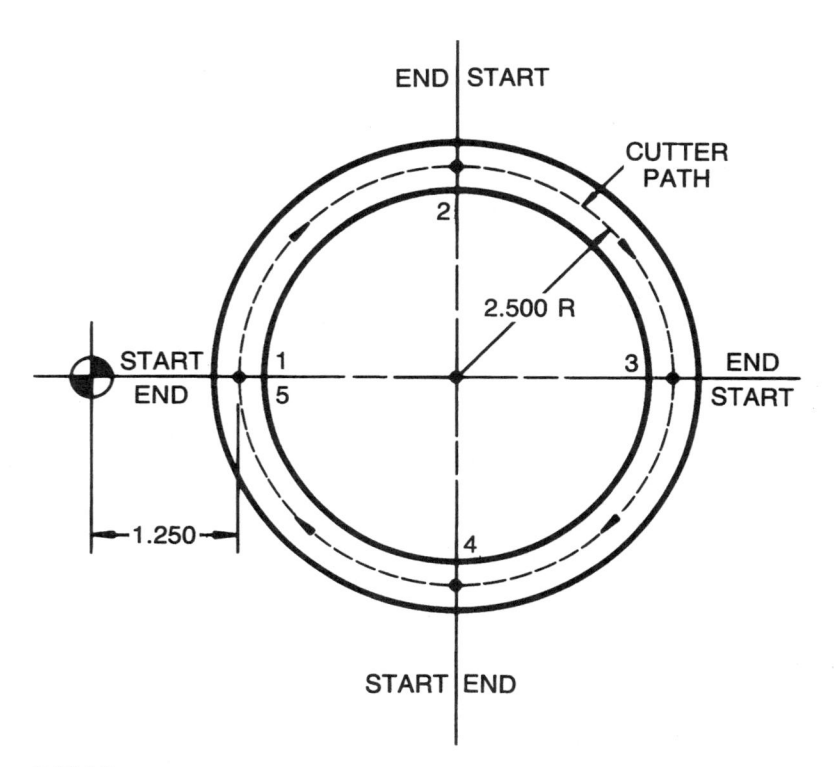

ANSWERS

6. (1) _____
   (2) _____
   (3) _____
   (4) _____
   (5) _____
   (6) _____
   (7) _____
   (8) _____
   (9) _____
   (10) _____
   (11) _____
   (12) _____
   (13) _____
   (14) _____
   (15) _____
   (16) _____
   (17) _____
   (18) _____

```
O4312
N5 G(1) G17 G20 G(2) G40
N10 T01 M(3)
N15 G(4) H01 Z1.0
N20 S4800 M(5)
N25 G00 X(6) Y0 Z.1
N30 G(7) Z(8) F3.0
N35 G02 X(9) Y(10) I(11) J0
N40 X6.25 Y(12) I0 J(13)
N45 X(14) Y-2.50 I-2.50 J(15)
N50 X1.25 Y0 I(16) J2.50
N55 G00 Z(17) G49 M05
N60 X(18)
N65 M30
%
```

Test #41 Circular Interpolation (Section 15, Unit 43)

7. Polar coordinates are used to quickly calculate and define __?__ and __?__ movements if the arc radius and angle are provided.

7. _____
   _____

8. Polar coordinates movements counterclockwise from the polar reference line are __?__, those clockwise are __?__.

8. _____
   _____

_____
Total—50 marks

# TEST #42
# SUBROUTINES OR MACROS (SECTION 15, UNIT 44)

Several types of program logic can be used to streamline CNC programming. Subroutines or macros make it easy to store and recall frequently used instruction sets. Logical loops provide a way to repeat a group of commands as often as necessary. Canned or fixed cycles are preset commands that cause the machine to perform specified cutting operations.

Place the correct word(s) in the space(s) at the right-hand side of the page to the statement complete and true.

ANSWER

1. __?__ logic is used when a program follows each program step in sequence from beginning to end.  1. _____

2. A miniprogram used within a main program to store frequently used instruction sequences is called a __?__ or a __?__.  2. _____  _____

3. A subroutine program is usually stored under a program number and is recalled into the main program by a __?__ statement.  3. _____

4. Sequence numbers of a subroutine should be __?__ enough so they don't conflict with the sequence numbers of the __?__ program.  4. _____  _____

5. The __?__ code is used to call a subroutine. The __?__ code is used at the end of a subroutine to return to the main program.  5. _____  _____

6. A __?__ in a program is designed to repeat a specific group of commands as often as they are required.  6. _____

7. A __?__ is a sequence of commands that, once started, repeats itself continuously until the machine is stopped; a __?__ repeats itself only a specified number of times.  7. _____  _____

8. The __?__ command causes the program to return to a certain line in the program to repeat the same commands.  8. _____

9. A __?__ is a preset combination of programmed commands that causes the machine or spindle to perform specified machining tasks.  9. _____

10. A canned cycle is __?__ into the controls, meaning that it cannot be erased.

11. A canned cycle shortens and simplifies the programming required for __?__ operations.

12. On most CNC lathes, the number of movements possible with the G84 turning-cycle code is __?__ .

13. Most CNC machines and control systems use the standard __?__ numbering system for canned or fixed cycles.

14. The __?__ cycle is used for counterboring and spot facing operations where a smooth surface is desired.

15. In a turning cycle, factors that would change from one part to another include the __?__ of the rough stock and the __?__ of cut taken for each pass.

10. _____

11. _____

12. _____

13. _____

14. _____

15. _____
    _____

_____

Total—20 marks

# TEST #43
# CNC MACHINING CENTERS (SECTION 15, UNIT 45)

CNC machining centers are widely used in industry because of their accuracy, reliability, repeatability, and productivity. Their main advantages to industry are increased machine uptime, maximum part accuracy, reduced scrap, less inspection time, reduced inventory, and lower manufacturing costs. The most common machine used in industry and training programs is the vertical machining center.

Fill in the missing program information, indicated by the numbers in brackets, in the correct spaces in the right-hand column.

*Program Notes*
- Program in the absolute mode starting at the XY zero (center of part), Fanuc compatible control.
- Material is aluminum (CS300), 1/2 x 4 1/8 x 4 1/8 in. part is held on a vacuum plate or holding fixture.
- Machine the edges to size, clockwise starting at **Point A**, with a 1 in. diameter 4-flute HSS end mill at 15 in. feed.
- Machine the hexagonal slot 1/4 in. wide x 1/4 in. deep, clockwise starting at **Point B**.
- Machine the circular slot 1/4 in. wide x 1/8 in. deep, clockwise starting at **Point C**.
- Machine the 4 cross patterns 1/4 in. wide x 3/4 in. long x 1/4 in. deep, clockwise starting at **Point D**.
- Drill the three 3/8 in. diameter holes through the part.

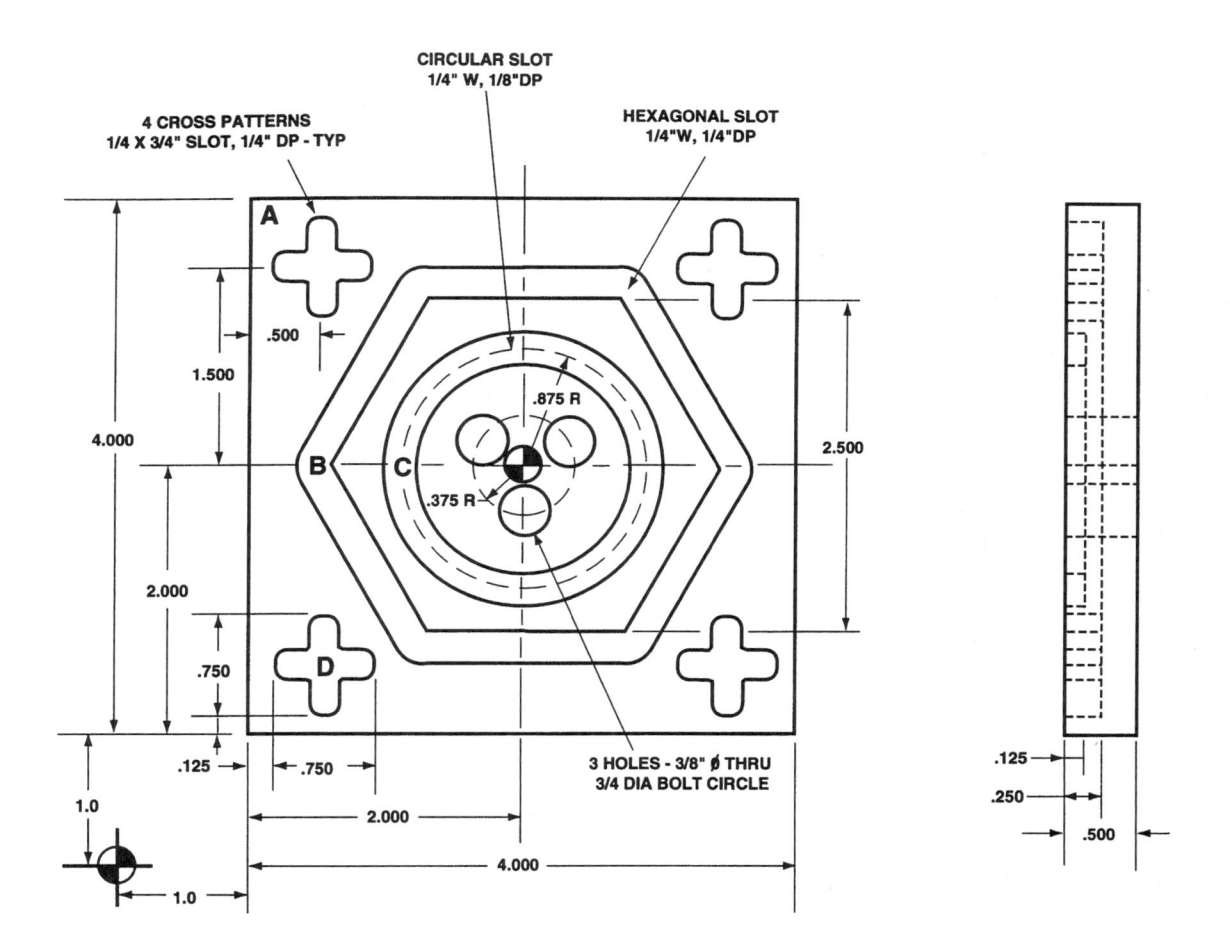

%
04300
N5 G54 G17 G20 G (1) G40
N10 T1 M06 (1.0 end mill)
N15 S1200 M(2)
N20 G43 H01 X-2.600 Y3.100 Z.1
N25 G01 Z(3) F15.0
N30 G41 D01 X-2.000 Y(4)
N35 X2.000
N40 Y(5)
N45 X-2.000
N50 Y(6)
N55 G40 X-2.500 Y3.10
N60 G00 Z1.0 G49 M(7)
N65 G00 T2 M(8) (.250 end mill)
N70 S(9) M03
N75 G43 H02 X-1.5877 Y(10) Z.1
N80 G01 Z(11) F15.0
N85 X-.7939 Y1.375
N90 X(12)
N95 X(13) Y0
N100 X.7939 Y(14)
N105 X(15)
N110 X-1.5877 Y(16)
N115 Z(17)
N120 G00 X-.875
N125 G(18) Z(19) F15.0
N130 G(20) X-.875 I(21)
N135 G01 Z(22)
N140 G(23) X-1.500 Y-1.500
N145 M98 P4301
N150 G00 X-1.500 Y(24) Z.1
N155 M(25) P4301
N160 G(26) X(27) Y1.500 Z.1
N165 M98 P(28)
N170 G00 X(29) Y-1.500 Z.1
N175 M98 P4301
N180 G00 Z(30) G49 M05
N185 G00 T(31) M06 (spot drill)
N190 S(32) M03
N195 G(33) H03 X0 Y-.375 Z.1
N200 G(34) G(35) Z-.219 R.1 F10.0

ANSWERS

1. _____
2. _____
3. _____
4. _____
5. _____
6. _____
7. _____
8. _____
9. _____
10. _____
11. _____
12. _____
13. _____
14. _____
15. _____
16. _____
17. _____
18. _____
19. _____
20. _____
21. _____
22. _____
23. _____
24. _____
25. _____
26. _____
27. _____
28. _____
29. _____
30. _____
31. _____
32. _____
33. _____
34. _____
35. _____
36. _____
37. _____
38. _____
39. _____
40. _____
41. _____

N205 X-.3248 Y.1875
N210 X(36)
N215 G(37) Z1.0 G(38) M05
N220 G00 T4 M(39) (.375 drill)
N225 S3200 M03
N230 G43 H04 X0 Y(40) Z.1
N235 G83 G99 Z(41) Q.210 R.1 F10.0
N240 X-.3248 Y(42)
N245 X(43)
N250 G00 Z1.0 G(44) M05
N255 G90 G00 X-3.0 Y2.0 Z1.0
N260 M(45)
%
SUBPROGRAM - P4301 Cross Patterns
%
O4301
N1000 G91 G00 X-.375 Z.1
N1005 G01 Z(46) F10.0
N1010 X(47)
N1015 Z(48)
N1020 G00 X-.375 Y(49)
N1025 G01 Z-.250
N1030 Y-.750
N1035 Z(50)
N1040 G90
N1045 M99
%

42. _____
43. _____
44. _____
45. _____
46. _____
47. _____
48. _____
49. _____
50. _____

# TEST #44
# TURNING CENTERS (SECTION 15, UNIT 46)

CNC turning centers are designed to machine shaft-type workpieces which may be held in a chuck at one end and supported by a tailstock center at the other end. Chucking centers are used for machining work held in a chuck or fixture. The combination turning/milling center can perform turning and some milling, drilling, or tapping operations on round work.

Fill in the missing program information, indicated by the numbers in brackets, in the correct spaces in the right-hand column.

*Program Notes*
- Program in the absolute mode, on a Fanuc compatible controller, starting at the XY zero at the right edge of the part.
- The tool reference point is X2 Y2.
- Use diameter programming to:
  (a) rough cut the form to within .060 in. of size.
  (b) finish cut the form to size.
- A cemented-carbide insert tool will be used for machining.
- Material: 1 1/4 in. diameter machine steel (CS 300), 3 in. long.

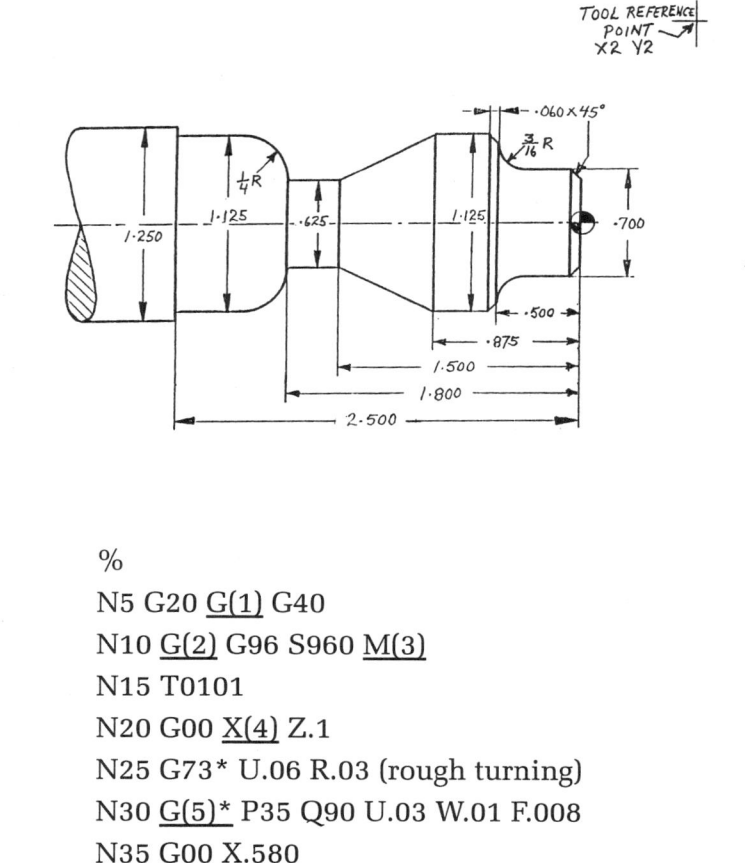

%
N5 G20 G(1) G40
N10 G(2) G96 S960 M(3)
N15 T0101
N20 G00 X(4) Z.1
N25 G73* U.06 R.03 (rough turning)
N30 G(5)* P35 Q90 U.03 W.01 F.008
N35 G00 X.580
N40 G01 Z(6)
N45 X.700 Z(7)
N50 Z-.3125
N55 G02 X(8) Z-.500 R(9)

ANSWERS
1. _____
2. _____
3. _____
4. _____
5. _____
6. _____
7. _____
8. _____
9. _____
10. _____
11. _____
12. _____
13. _____
14. _____
15. _____
16. _____
17. _____
18. _____
19. _____
20. _____
21. _____

Test #44 Turning Centers (Section 15, Unit 46)

N60 G(10) X(11) Z-.560
N65 Z(12)
N70 X(13) Z-1.500
N75 Z(14)
N80 G(15) X1.125 Z(16) R(17)
N85 G(18) Z(19)
N90 X(20)
N95 G00 X(21) Z.1
N100 G(22) * P35 Q(23) F.005 (finish turning)
N105 G00 X(24) Y2.0 M05
N110 M(25)
%
*Rough and finish turning codes;
teaching-size machines.

22. _____
23. _____
24. _____
25. _____

_____
Total—25 marks

# REVIEW TEST #8
# CNC OPERATIONS (SECTION 15, UNITS 39 TO 46)

## PART 1

Consider whether each statement is true or false, then indicate your choice by circling the proper letter in the right-hand column.

|     |                                                                                                                                                      |     | ANSWER |   |
| --- | ---------------------------------------------------------------------------------------------------------------------------------------------------- | --- | ------ | - |
| 1.  | Program preparation and setup for CNC machines is usually a lengthy process.                                                                         | 1.  | T      | F |
| 2.  | Because of the expense of a CNC system, the operator usually makes more trial measurements and trial cuts.                                           | 2.  | T      | F |
| 3.  | The two types of programming modes used in CNC systems are the absolute mode and the incremental mode.                                               | 3.  | T      | F |
| 4.  | Nonmodal codes stay in effect in the program until they are changed by another code.                                                                 | 4.  | T      | F |
| 5.  | In contouring or continuous-path CNC operations, the cutting tool is not in contact with the workpiece while moving from one coordinate position to the next position. | 5.  | T      | F |
| 6.  | Tool settings and offsets allow the programmer or operator to meet unexpected tooling problems.                                                      | 6.  | T      | F |
| 7.  | The programming code letter T refers to the time of movement of the tool.                                                                            | 7.  | T      | F |
| 8.  | The command X - .875 tells the tool to move .875 above part zero.                                                                                    | 8.  | T      | F |
| 9.  | Two requirements that the programmer must consider for contouring are machine feed rates and tool geometry.                                          | 9.  | T      | F |
| 10. | One common method of programming an arc is called radius programming.                                                                                | 10. | T      | F |
| 11. | A subroutine is a group of instructions that can be recalled from memory as a group to solve recurring problems.                                     | 11. | T      | F |
| 12. | Dumb loops are rarely used in CNC programming.                                                                                                       | 12. | T      | F |
| 13. | When a workpiece is set up, it should be loosely fastened in order to minimize the stress of machine startup.                                        | 13. | T      | F |
| 14. | Whenever possible, use only one clamp to hold a part so that vibration will be minimized.                                                            | 14. | T      | F |

15. The headstock of CNC chucking and turning centers contains the driving mechanism for the machine spindle.  15. T F

16. Self-centering chucks generally have higher gripping forces and are more accurate than other chuck types.  16. T F

## PART 2

Place the correct word(s) in the space(s) at the right-hand side of the page to make the statement complete and true.

ANSWER

17. Coded instructions used in numerical control consist of __?__ , __?__ of the alphabet, and __?__ .  17. _____
_____
_____

18. __?__ address is the most common type of programming format used for CNC programming systems.  18. _____

19. CNC information is generally programmed in blocks of __?__ words.  19. _____

20. The system of __?__ coordinates is very important to the successful operation of CNC machines.  20. _____

21. A milling cut must be programmed at a feed __?__ to suit the material being cut.  21. _____

22. When performing __?__ milling, it is important that the machine table be equipped with a backlash eliminator.  22. _____

23. The cutting speed for milling cutters is expressed in either __?__ or __?__ per minute.  23. _____
_____

24. __?__ interpolation was developed to simplify the CNC programming.  24. _____

25. Every CNC machine tool has a basic __?__ point that is usually the centerline of the spindle and the end of the spindle nose.  25. _____

26. The offsets for cutting-tool geometry must always be made __?__ to the surface being machined, regardless of the __?__ that the surface faces.  26. _____
_____

27. A subroutine must be written as a separate __?__ .  27. _____

28. The GOTO command causes the program to return to a certain __?__ in the program in order to __?__ the same commands.

28. _____
    _____

29. The __?__ is the part of a CNC machining center that prevents distortion and provides vertical movement in the direction of the Z- axis.

29. _____

30. Tool changers are usually bi-directional and take the __?__ travel distance to randomly access a tool.

30. _____

31. The __?__ is the mechanism that holds tools for external and internal operations in a turning center.

31. _____

32. __?__ force increases as the speed of rotation increases and tries to move the chuck jaws outward.

32. _____

## PART 3

Select the correct answer for each question and circle the letter in the right-hand column indicating your choice.

                                                                    ANSWER

33. In a computer, the internal memory where data is stored permanently and can only be read out is called the
    (A) RAM          (C) ROM
    (B) DASD         (D) microprocessor

33.   A    B    C    D

34. One of the main features of CNC is its ability to
    (A) compute          (C) think mechanically
    (B) evaluate         (D) judge

34.   A    B    C    D

35. When the spindle moves rapidly from one location to another on the workpiece, it
    (A) runs          (C) roughs
    (B) reams         (D) rapids

35.   A    B    C    D

36. In the manual data-input mode, data is entered into the machine control unit (MCU) through the

   (A) mode control switch
   (B) central processing unit
   (C) keyboard
   (D) drive motor

   36.  A   B   C   D

37. Dimensioning for both contouring and positioning uses

   (A) polar coordinates
   (B) rectangular coordinates
   (C) arc coordinates
   (D) none of the above

   37.  A   B   C   D

38. In circular interpolation, many curves and free-form shapes can be closely approximated with a series of

   (A) spirals
   (B) arcs
   (C) circles
   (D) cubic shapes

   38.  A   B   C   D

39. In conventional milling, the cutter rotation and the table feed are moving in

   (A) opposite directions
   (B) 90-degree angles
   (C) circles
   (D) 45-degree angles

   39.  A   B   C   D

40. In a continuous-path control system, the drive motors of the X and Y axes can operate at different

   (A) voltages
   (B) rates of cut
   (C) frequencies
   (D) rates of speed

   40.  A   B   C   D

41. The method used to move contouring machines from one point to the next is called

   (A) circular programming
   (B) path extension
   (C) offsetting
   (D) interpolation

   41.  A   B   C   D

42. The factor that a programmer does NOT have to consider in contour programming is

   (A) tool lubrication
   (B) type of cutting tool
   (C) feed rate
   (D) depth of cut

   42.  A   B   C   D

43. Loops can save valuable programming time because they require only one

   (A) programmer
   (B) word
   (C) statement
   (D) cycle

   43.  A   B   C   D

44. The tool that is used to accurately spot hole locations for a drill that is to follow is called a

   (A) spade drill
   (B) center drill
   (C) stub drill
   (D) shell drill

   44.  A   B   C   D

45. The basic difference between CNC bench-top lathes and conventional CNC turning centers is the position of the

   (A) threads
   (B) slides
   (C) chuck
   (D) cutting tools

   45.  A   B   C   D

Total—50 marks

# TEST #45
# HEAT TREATMENT OF STEEL (SECTION 16, UNIT 47)

Understanding the properties and heat treatment for various steels is important to the proper functioning of a part. The selection of the proper steel and its heat treatment will ensure that the manufactured part will perform properly in use.

Place the correct word(s) in the space(s) at the right-hand side of the page to make the statement complete and true.

ANSWER

1. Heat treatment changes the __?__ of steel.                                             1. _____

2. Three qualities obtained through heat treatment are __?__,                             2. _____
   __?__, and __?__ resistance.                                                              _____
                                                                                              _____

3. The main elements in steel are __?__ and __?__.                                        3. _____
                                                                                              _____

4. Low carbon steel cannot be hardened throughout, but its                                 4. _____
   surface can be __?__ hardened.

5. High-speed steel cutting tools provide good wear __?__ and                             5. _____
   __?__ hardness.                                                                            _____

6. Two standard steel identifications are __?__ and __?__.                                6. _____
                                                                                              _____

7. Pearlite is a saturated mixture of __?__ and __?__.                                    7. _____
                                                                                              _____

8. __?__ __?__ point is the lowest temperature at which steel                             8. _____
   may be quenched in order to harden it.                                                     _____

9. The operation of __?__ removes the brittleness and                                     9. _____
   toughness of the steel.

10. __?__ improves the grain structure and removes stress and                             10. _____
    strains.

11. __?__ is a process that improves the machinability of the                             11. _____
    metal.

12. __?__ is the process of heating metal uniformly to its proper                         12. _____
    __?__ and then quickly quenching or cooling it in the proper                              _____
    medium.

13. Pack hardening is also called __?__ .  13. _____
14. __?__ hardening is used to harden ways on machine tools.  14. _____
15. Always wear a __?__ shield, __?__ , and protective __?__ when working with hot metal.  15. _____
_____
_____

_____

Total—25 marks

# TEST #46
# ARTIFICIAL INTELLIGENCE (SECTION 17, UNIT 48)

The use of artificial intelligence enables machines to perform functions that normally require human understanding. It requires the use of a body of expert knowledge and specialized computer hardware and software. Artificial intelligence systems can imitate human sensing systems and decision-making processes.

Place the correct word(s) in the space(s) at the right-hand side of the page to make the statement complete and true.

ANSWER

1. Artificial intelligence requires a large __?__ , which is developed from a body of knowledge about the particular subject.

   1. _____

2. Programming languages developed for artificial intelligence need to handle __?__ rather than __?__ .

   2. _____
      _____

3. Artificial intelligence uses __?__ logic to deal with the uncertainties of continuously changing production problems.

   3. _____

4. In routine operations, artificial intelligence uses knowledge-based __?__ rules to make decisions.

   4. _____

5. Artificial intelligence systems use processes similar to the methods used by the human __?__ to solve problems.

   5. _____

6. Although artificial intelligence can perform many human functions, it does not have the ability to __?__ .

   6. _____

7. __?__ systems provide decision-making capabilities for problems that require specialized knowledge.

   7. _____

8. Special systems can simulate one or more of the human __?__ , such as sight, hearing, touch, and smell.

   8. _____

9. A __?__ is a mechanism or device that uses adaptive control, sensing, and learning to perform physical tasks.

   9. _____

10. __?__ programming uses symbols to represent objects and relationships.

    10. _____

11. __?__ refers to the process of using rules from past experiences to guide present thinking.

    11. _____

12. Stored knowledge is used in the problem-solving process to __?__ alternatives and __?__ facts.

12. _____
_____

13. As product complexity increases, so does the __?__ operation that ensures that the product will perform properly.

13. _____

14. __?__ systems can analyze and translate two-dimensional camera images into three-dimensional computer images.

14. _____

15. Dimensional gaging systems generally can include the following categories: __?__ lasers, __?__ cameras, and __?__ array systems.

15. _____
_____
_____

16. In a triangulation-type gaging system, __?__ sensors are used to accurately measure the distance between the probe and the surface of the part.

16. _____

_____

Total—20 marks

# TEST #47
# COMPUTER MANUFACTURING TECHNOLOGIES
# (SECTION 17, UNIT 49)

The increased use of computers in manufacturing is producing immediate and widespread benefits. Major technical applications include design and drafting, planning and control, and integration of functions. Advantages include more efficient information usage, fewer errors, and improved communications.

Consider whether each statement is true or false, then indicate your choice by circling the proper letter in the right-hand column.

|     |     | ANSWER |     |
| --- | --- | --- | --- |
| 1. | The earliest uses of computers in manufacturing were in engineering applications. | 1. T | F |
| 2. | The principal function of a CAD system is to produce engineering drawings. | 2. T | F |
| 3. | Because of the computer's abilities, it is generally not important for the CAD computer program to suit or match the purpose for which it will be used. | 3. T | F |
| 4. | The main operating components of a CAD system are the interactive graphics terminal, digitizer, and the computer. | 4. T | F |
| 5. | Generally, images on the CRT screen can be rotated, inverted, enlarged, and examined in close-ups. | 5. T | F |
| 6. | Terminals equipped for digitizing are usually connected to a sensitized table where a drawing is traced. | 6. T | F |
| 7. | The computer in a CAD system usually performs one main function. | 7. T | F |
| 8. | CAD relates to the functions involved in producing a finished product; CAM relates to the functions involved in getting ready to produce a product. | 8. T | F |
| 9. | Even with low-cost CAM systems, it is generally not possible for small machine shops to develop new products quickly. | 9. T | F |
| 10. | The most important component of a CAD/CAM system is its CRT screen and keyboard. | 10. T | F |

Select the correct answer for each question and indicate your choice by circling correct letter in the right-hand column.

|     |                                                                                                  |     |   |   | ANSWER |   |
|-----|--------------------------------------------------------------------------------------------------|-----|---|---|--------|---|
| 11. | All part designs begin with<br>(A) a model (C) a concept<br>(B) a bill of materials (D) a digitizer | 11. | A | B | C | D |
| 12. | Canned software routines that make it easy to perform common machining operations quickly are<br>(A) databases (C) CAM calculators<br>(B) plotters (D) macros | 12. | A | B | C | D |
| 13. | The main reason for implementing CAM may be the need to<br>(A) be competitive (C) save shop space<br>(B) lower inventories (D) share data | 13. | A | B | C | D |
| 14. | Which of the following is not one of the inputs to productivity?<br>(A) labor (C) capital<br>(B) inventory (D) raw material | 14. | A | B | C | D |
| 15. | The main purpose of CIM is to<br>(A) produce engineering drawings (C) provide a parts analysis<br>(B) link automated functions (D) coordinate part measurements | 15. | A | B | C | D |
| 16. | A centralized CIM database allows product information to be<br>(A) made obsolete (C) flexible<br>(B) conceptualized (D) reused | 16. | A | B | C | D |

17. Which of the following types of computer system is not one of the three levels of computers that make up a CIM system?

    (A) floor level  
    (B) CAD/CAM  
    (C) mainframe  
    (D) micro level

17.  A  B  C  D

18. Desktop workstations allow engineers and designers to use and expand the central

    (A) database  
    (B) machine records  
    (C) inventory  
    (D) processing

18.  A  B  C  D

19. Using common computer data produces

    (A) higher-quality products  
    (B) shorter cycle times  
    (C) faster design changes  
    (D) all of the above

19.  A  B  C  D

20. Which of the following benefits is not a result of using CIM?

    (A) increased product lead time  
    (B) reduced labor costs  
    (C) reduced design costs  
    (D) none of the above

20.  A  B  C  D

Total—20 marks

# TEST #48
# COORDINATE MEASURING SYSTEMS (SECTION 17, UNIT 50)

Coordinate measuring systems can quickly and accurately measure and inspect parts produced by CNC machine tools. They keep pace with high production rates and make measurements at various stages of the manufacturing process. benefits include reduced downtime, lower scrap rates, and higher productivity.

Place the correct word(s) in the space(s) at the right-hand side of the page to make the statement complete and true.

ANSWER

1. Coordinate measuring systems are used to measure __?__ and __?__ dimensions.

2. Inspectors using conventional tools cannot keep up with the __?__ of CNC machines.

3. A __?__ is an advanced, multi-purpose system used to quickly inspect machined parts.

4. A CMM is typically used for inspecting the __?__ part produced by a CNC machine and for random __?__ of parts during a production run.

5. A CMM consists of an indicator probe supported on three perpendicular __?__ .

6. The most common measuring system used on CMMs is the __?__ scale.

7. Ultra-accurate CMMs usually have a fiber-optic __?__ measuring system for greater __?__ .

8. The digital readout of the CMM provides accurate, instant information about the measuring probe's __?__ and __?__ .

9. The structural design of a vertical CMM ensures that the machine will be able to precisely __?__ measurements time after time.

10. __?__ CMMs are designed for use with large, bulky workpieces such as gear cases and engine blocks.

11. A CMM controls quality __?__ a manufacturing process rather than after it.

12. A CMM can supply data on factors such as tool __?__ and part __?__.

13. A CMM saves inspection time by eliminating multiple __?__.

14. Because it is flexible, a CMM can be used in a production line that __?__ product frequently.

15. __?__ sampling of product helps to reduce inspection time and to identify potential problems.

12. _____
    _____
13. _____
14. _____
15. _____

_____

Total—20 marks

# TEST #49
# ELECTRICAL DISCHARGE MACHINING (EDM)
# (SECTION 17, UNIT 51)

Electrical discharge machining (EDM) removes metal from parts by the action of a short-duration electrical discharge. It can be used to produce a wide variety of shapes and to cut materials that have already been hardened. The process is controlled automatically, and it can be used to cut any material that is electrically conductive.

Place the correct word(s) in the space(s) at the right-hand side of the page to make the statement complete and true.

ANSWER

1. In EDM, metal is removed by electric __?__ erosion.                                  1. _____

2. The __?__ type of EDM uses a cutting tool that is shaped to                           2. _____
   the form of the cavity. The tool or electrode __?__ its form in                         _____
   the workpiece.

3. The __?__ type of EDM uses a vertical traveling wire that                             3. _____
   follows a horizontal path through the workpiece.

4. In ram-type EDM, the dielectric fluid acts as an electrical                           4. _____
   __?__ ; it also __?__ the electrode and workpiece so they                                _____
   don't get too hot and it __?__ the metal particles out of the                            _____
   gap.

5. In order to be effective, the dielectric fluid must be                                5. _____
   circulated under constant __?__ .

6. The __?__ controls the arc gap distance between the                                   6. _____
   electrode and the workpiece.

7. Metal-removal rates for EDM are __?__ than for conventional                           7. _____
   machining methods; the amount of metal removed with each                                 _____
   pulse is directly related to the __?__ of the pulse.

8. Graphite electrodes produce __?__ metal-removal rates than                            8. _____
   other electrodes; the harder the material, the __?__ the metal-                          _____
   removal rate.

9. Higher metal-removal rates result in __?__ surface finishes.                          9. _____

10. In wire-cut EDM, the moving wire never __?__ the                                     10. _____
    workpiece.

11. Effective wire electrodes need to be good __?__ of electricity and heat, and should have a high __?__ strength and a high __?__ point.

12. Untreated __?__ water is not suitable for use as the dielectric fluid in EDM wire operations.

13. In order for any material to be cut by EDM, it must be able to conduct __?__ .

11. _____
    _____
    _____

12. _____

13. _____

_____

Total—20 marks

# TEST #50
# FLEXIBLE MANUFACTURING SYSTEMS (SECTION 17, UNIT 52)

Flexible manufacturing systems (FMS) allow producers to respond quickly to changing customer needs. Manufacturers can produce a variety of parts in the required volumes, and they can do it while minimizing time lost in equipment setups and changes.

Consider whether each statement is true or false, then indicate your choice by circling the proper letter in the right-hand column.

|  |  |  | ANSWER |
|---|---|---|---|
| 1. | A typical FMS is designed to provide maximum efficiency in producing one type of part. | 1. | T  F |
| 2. | FMS often uses equipment during off-hours, such as during second and third shifts. | 2. | T  F |
| 3. | The greatest need for FMS is in situations calling for high volumes of parts and high levels of part variety. | 3. | T  F |
| 4. | Small parts require more automated methods for systems work than large parts. | 4. | T  F |
| 5. | One factor that often determines the manufacturing method to be used is the required workpiece accuracy. | 5. | T  F |
| 6. | The shorter the life cycle of the product design, the less desirable it is to have a flexible system that permits rapid changes. | 6. | T  F |
| 7. | In most manufacturing operations, there is a conflict between flexibility and productivity. | 7. | T  F |
| 8. | The level of machine specialization is not as flexible as the level of process specialization. | 8. | T  F |
| 9. | High-volume production almost always results in high efficiency. | 9. | T  F |
| 10. | Specialized, single-function machine tools are usually a part of a modern FMS. | 10. | T  F |
| 11. | Companies are finding it more difficult to invest in flexible automation because of the availability of better graphics and computer software for product design. | 11. | T  F |
| 12. | A flexible system to meet changing production needs is usually designed with a modular approach. | 12. | T  F |
| 13. | A stand-alone machining center usually works on one type of part at a time. | 13. | T  F |

14. A manufacturing cell is designed to complete all the operations on a part before that part leaves the cell.     14. T   F

15. A palletized cell is generally used in low-variety, high-volume production situations.     15. T   F

16. A robot cell is typically used in high-variety, low-volume production situations.     16. T   F

17. From an economic standpoint, it is best to run FMS for 24 hours a day, not for shorter periods.     17. T   F

18. Factories of the future will produce a specialized, narrow range of products and will require little human intervention.     18. T   F

19. Tool management refers to the system of purchasing tools at the lowest cost for the manufacturing operation.     19. T   F

20. Material-handling systems are generally considered to be an important part of FMS operations.     20. T   F

Total—20 mark

# TEST #51
# GROUP TECHNOLOGY (SECTION 17, UNIT 53)

Group technology refers to classifying parts into various part families based on specified characteristics. All parts within a given family will require similar machining operations. Machines can then be grouped efficiently to process the various part families.

Select the correct answer for each question and indicate your choice by circling the correct letter in the right-hand column.

ANSWER

1. In addition to classifying parts, group technology also involves the clustering of machines into

    (A) inventory groups  (C) cells
    (B) engineering units  (D) families

    1.  A  B  C  D

2. Which of the following factors increases when parts are grouped according to their similarities?

    (A) setup time  (C) overall production time
    (B) work-in-process inventory  (D) productivity

    2.  A  B  C  D

3. When workers are able to see their contributions to the final product, this increases their

    (A) confidence  (C) throughput time
    (B) job satisfaction  (D) judgment

    3.  A  B  C  D

4. The group technology coding system identifies the similarities in parts according to key

    (A) shop locations  (C) customer orders
    (B) manufacturing features  (D) raw-material requirements

    4.  A  B  C  D

5. The first digit of a numeric part code generally identifies the

    (A) heat treatment history  (C) shop machining location
    (B) machining sequence  (D) major family classification

    5.  A  B  C  D

Test #51 Group Technology (Section 17, Unit 53)

6. Which of the following is not a benefit of a well-designed parts classification system?   6.   A   B   C   D

   (A) increased throughput tune
   (B) fewer design duplications
   (C) quick retrieval of information
   (D) improved production planning

7. Which of the following part-code categories would refer to the design of a part?   7.   A   B   C   D

   (A) setup time
   (B) cutting tools
   (C) workholding devices
   (D) part material

8. Which of the following part-code categories would refer to the manufacture of a part?   8.   A   B   C   D

   (A) main shape
   (B) tolerance
   (C) machine tool
   (D) functions

9. After a major part family has been defined, the next step in coding is to break the family into   9.   A   B   C   D

   (A) inventory groups
   (B) larger groups
   (C) smaller groups
   (D) customer groups

10. One of the main goals of group technology is to eliminate   10.   A   B   C   D

    (A) excess workers and machines
    (B) finished goods inventory
    (C) waste and duplication
    (D) database updates

Place the correct word(s) in the space(s) at the right-hand side of the page to make the statement complete and true.

ANSWER

11. The purpose of __?__ is the efficient distribution of the workload among the available machine tools; this requires information about the __?__ of parts to be purchased.   11. _____
   _____

12. It is important to avoid designing a __?__ part if a part is already available in a part family.   12. _____

13. A design retrieval system eliminates the __?__ of previous designs stored in a database and can reduce design costs.

14. The two types of Computer-Aided Process Planning (CAPP) systems are the __?__ system and the __?__ system.

15. In cellular manufacturing, the cells are designed to handle all the __?__ required on a family of parts.

16. Whenever possible, any job assigned to a cell should be __?__ within that cell.

17. In a shop's data-processing operations, __?__ date describes the part to be designed and manufactured; __?__ data describes the manufacturing operations and procedures required to make the part.

13. _____

14. _____
    _____

15. _____

16. _____

17. _____
    _____

_____

Total—20 marks

# TEST #52
# JUST-IN-TIME MANUFACTURING (SECTION 17, UNIT 54)

The goals of Just-In-Time (JIT) manufacturing include improving productivity, reducing waste, and lowering costs. Deliveries of materials, movements of materials, and availability of machines, tools, and labor are all precisely planned and timed to achieve these goals.

Consider whether each statement is true or false, then indicate your choice by circling the proper letter in the right-hand column

ANSWER

1. When JIT was first begun, the concept was to supply or produce parts only when they were required for the manufacturing process.     1. T F

2. In JIT, producing more than is actually required is desirable because unexpected delays or problems could arise.     2. T F

3. Moving materials from one place to another in a shop is a costly waste of time.     3. T F

4. The more inventory you can produce and keep on hand, the more efficiently you are running the operation.     4. T F

5. In a machine shop, any motion that does not add value to the product is wasteful.     5. T F

6. In JIT, it is permissable to have a certain percentage of defective parts because of the speed of the operations.     6. T F

7. Typically, JIT results in an increase in the manufacturing space required to produce the necessary parts.     7. T F

8. In a JIT system, the production time usually increases at first and then decreases and levels out.     8. T F

9. Most companies that use JIT find that the main benefit at the start of the program is a reduction in product defects.     9. T F

10. Inventories tie up company money that could be used for other purposes.     10. T F

11. The difficulties in making a JIT system work are usually the result of the machinists' resistance to change.     11. T F

12. Raw material that arrives late holds up production, while material that arrives early is desirable because you can hold it until you need it.     12. T F

13. In a JIT system, the number of suppliers for a given product is usually reduced to a very few.     13.   T   F

14. In order to ensure product quality, most JIT companies try to reduce their reliance on outside suppliers for design input.     14.   T   F

15. The typical JIT deliveries from a supplier are large orders with flexible order schedules.     15.   T   F

16. One goal of JIT purchasing is to completely eliminate the inspection of incoming goods from suppliers.     16.   T   F

17. One advantage of a JIT system is that it can usually be up and running in a short time, so positive results come quickly.     17.   T   F

18. When a JIT system is installed, it usually affects the management levels of the organization more than the shop level.     18.   T   F

19. After JIT is installed, the different departments in the company are not as dependent upon each other as they were before.     19.   T   F

20. Although work-in-process inventories generally rise under JIT systems, this is more than offset by reductions in raw materials and finished goods inventories.     20.   T   F

Total—20 marks

# REVIEW TEST #9
# MANUFACTURING TECHNOLOGIES (FIRST HALF)
# (SECTION 17, UNITS 48 TO 54)

## PART 1

Consider whether each statement is true or false, then indicate your choice by circling the proper letter in the right-hand column.

ANSWER

1. Artificial intelligence (AI) can perform many human tasks, but it cannot create.     1. T   F

2. The term heuristics refers to the use of past experiences to guide current thinking processes.     2. T   F

3. The major drawback of today's CAD systems is that a change made on one view of a part cannot be automatically included in other views.     3. T   F

4. One of the benefits of CAM systems is higher inventory levels, which allow the manufacturer to quickly fill customer orders.     4. T   F

5. In general, inspectors using conventional measuring tools have been able to keep up with the increased output of CNC machines.     5. T   F

6. The most common measuring system used on CMMs is the reference scale.     6. T   F

7. In ram EDM, the workpiece and the tool (electrode) must be placed in a working position so that they are in contact with each other.     7. T   F

8. One of the purposes of dielectric fluid is to flush metal particles out of the spark erosion area.     8. T   F

9. In flexible manufacturing systems, large parts will generally need more automated methods for systems work than smaller parts.     9. T   F

10. The longer the product-design life cycle, the more desirable it is to use less-flexible approaches to manufacturing.     10. T   F

11. The key to applying group technology principles is the efficient way that machines are grouped into cells.     11. T   F

12. Because of the need to maintain their trade secrets, most companies using JIT systems do not rely on suppliers for help in product design.

12.  T  F

**PART 2**

Place the correct word(s) in the space(s) at the right-hand side of the page to make the statement complete and true.

ANSWER

13. Expert systems use __?__ programming techniques to simplify software development and reduce the time and effort required.

13. _____

14. As the complexity of a product increases, so does the complexity of the __?__ operation to ensure that the product performs well.

14. _____

15. CAM generates computer data for the machining/manufacturing processes from the __?__ program.

15. _____

16. It is important that CAM software is compatible with the machine __?__ it will be used on.

16. _____

17. A CMM is a versatile machine that can measure almost any __?__ on a part at any stage of production.

17. _____

18. A ring bridge on a vertical CMM ensures that the machine will be stable and able to precisely __?__ measurements.

18. _____

19. One of the main benefits of EDM is that a part can be cut to the proper form while in its __?__ state.

19. _____

20. The __?__ method of circulating dielectric fluid is especially valuable for very small holes and deep holes.

20. _____

21. A series of parts must often be produced in a one-to-another relationship because they are assembled as __?__ unit.

21. _____

22. The __?__ cell is generally used for high-variety, low-volume production items; it does not have central computer control.

22. _____

23. The efficient distribution of the production workload among the available machine tools is called load __?__ .

23. _____

24. In group technology, some machines within a cell should be __?__ enough to perform multiple operations.

24. _____

25. The basic objective of JIT manufacturing is the elimination of all __?__ in the operations.

25. _____

26. The number of __?__ for any particular product in JIT is reduced to a very few.

26. _____

## PART 3

Select the correct answer for each question and indicate your choice by circling the correct letter in the right-hand column.

ANSWER

27. Artificial intelligence can deal with continuously changing production problems by using fuzzy
    (A) thinking
    (B) logic
    (C) planning
    (D) creativity

27. A  B  C  D

28. The most important factor that produces quality products and increases productivity is
    (A) management
    (B) part families
    (C) technology
    (D) measurement systems

28. A  B  C  D

29. A CMM can measure almost any part, regardless of its
    (A) surface finish
    (B) shape
    (C) lubrication
    (D) defects

29. A  B  C  D

30. In EDM, the dielectric fluid keeps the electrode and the workpiece from becoming dangerously
    (A) hot
    (C) ionized
    (C) charged
    (D) hard

30. A  B  C  D

Review Test #9 Manufacturing Technologies (First Half) (Section 17, Units 48 to 54)

31. FMS brings flexibility and responsiveness to manufacturing to produce a part when it is required by the

　　(A) schedule
　　(B) inventory plan
　　(C) supplier
　　(D) market

31.　A　B　C　D

32. Two important functions in group technology classification and coding are part design retrieval and

　　(A) planning
　　(B) tool listings
　　(C) duplication
　　(D) standardization

32.　A　B　C　D

33. Implementation of JIT depends on the supplier's

　　(A) costs
　　(B) inventory layout
　　(C) flexibility
　　(D) profitability

33.　A　B　C　D

Total—33 marks

# TEST #53
# LASERS (SECTION 17, UNIT 55)

Laser systems are finding more uses in highly competitive industries such as automobiles and computer components. Because of their versatility, they are used in material processing, surveying, identification and tracking, communications, and precision measurement.

Consider whether each statement is true or false, then indicate your choice by circling the proper letter in the right-hand column.

|  |  |  | ANSWER |
|---|---|---|---|
| 1. | In the shop, laser systems are generally used for material processing, such as cutting or welding, and precision measuring. | 1. | T   F |
| 2. | A laser beam generally cannot be controlled over a wide range of temperatures at the point of focus. | 2. | T   F |
| 3. | The most common lasers are the solid, gas, and plasma types. | 3. | T   F |
| 4. | A laser uses two totally reflecting mirrors to bounce photons back and forth within the unit. | 4. | T   F |
| 5. | In a gas laser, mirrors allow the length of the optical path to be increased without increasing the overall length of the laser. | 5. | T   F |
| 6. | Gas lasers are usually slower than conventional machining when cutting sheet metal. | 6. | T   F |
| 7. | Excimer lasers emit ultraviolet light and remove material by erosion. | 7. | T   F |
| 8. | Noble gases are used in excimer lasers because they combine easily with other elements. | 8. | T   F |
| 9. | The excimer laser's biggest advantage is its ability to drill holes quickly. | 9. | T   F |
| 10. | The excimer process consists of high-resolution imaging and erosion. | 10. | T   F |
| 11. | Liquid lasers are best suited to machining thinner metal parts. | 11. | T   F |
| 12. | Lasercaving is especially effective in machining nonconductive materials such as ceramics and composites. | 12. | T   F |
| 13. | One of the advantages of lasercaving is its high material-removal rate. | 13. | T   F |

14. Gas discharge lasers are used to make accurate measurements of parts.     14. T   F

15. Laser welding, although very accurate, is generally slower than conventional welding.     15. T   F

16. In heat-treating operations, lasers reduce distortion of the workpiece.     16. T   F

17. Lasers are able to perform 100% inspection with 100% accuracy.     17. T   F

18. Lasers used in nondestructive testing can detect cracks that are invisible to the eye.     18. T   F

19. Lasers are used to mark and code small electronic components because they are much faster than other methods.     19. T   F

20. Bar coding is a data collection system used for marking, tracking, and inventory purposes.     20. T   F

Total—20 marks

# TEST #54
# ROBOTICS (SECTION 17, UNIT 56)

The ideal industrial robot is capable of performing many different tasks and operations. It usually consists of a single-arm device with a pair of grippers that are used to move something or perform a task. Robots are used in many applications where jobs are monotonous, physically difficult, or hazardous for human workers.

Place the correct word(s) in the space(s) at the right-hand side of the page to make the statement complete and true.

ANSWER

1. Industrial robots must be adaptable for many __?__, easily __?__ by an operator, and __?__ to operate so injuries are avoided.

    1. _____
    _____
    _____

2. The two main manufacturing areas where robots are used are in __?__ applications and __?__ applications.

    2. _____
    _____

3. An industrial robot must be able to reach all __?__ in the work area with speed and fluid motion.

    3. _____

4. Special electric robots operate in __?__ environments that would be dangerous for humans.

    4. _____

5. Robots used for moving tools, materials, and parts are often guided along __?__ embedded in the floors of factories.

    5. _____

6. __?__ are the parts of a robot that are usually designed for holding a paint gun or a spot welder.

    6. _____

7. In order for a robot to feel or fit parts properly, touch __?__ are located in the grippers or in the wrist.

    7. _____

8. Closed-circuit, black-and-white video __?__ are used in robot vision.

    8. _____

9. A robot's video system relies on one of two types of light sources, __?__ or __?__.

    9. _____
    _____

10. In the __?__ mode of operation, a robot can be taught by an operator who leads the robot arm through the required movements.

    10. _____

11. In the __?__ mode of operation, the robot's movements are controlled by a computer program.

    11. _____

Test #54 Robotics (Section 17, Unit 56)    171

12. Robots must be handled carefully in order to prevent __?__ .  12. _____

13. The robot working area should be enclosed by some form of __?__ to keep people from entering the area while the robot is in operation.  13. _____

14. Emergency stop buttons stop a robot by cutting off its __?__ .  14. _____

15. Computer-controlled robots are used to spot- __?__ car bodies.  15. _____

16. __?__ operations, which are performed over and over, are ideal applications for robots.  16. _____

_____

Total—20 marks

# TEST #55
# STATISTICAL PROCESS CONTROL (SECTION 17, UNIT 57)

Good quality control systems are necessary if a company wants to consistently produce quality goods that meet customer needs. Statistical quality control uses data from measuring instruments to produce control charts that are used to identify problem areas.

Select the correct answer for each question and indicate your choice by circling the correct letter in the right-hand column

ANSWER

1. The key to quality improvement rests with          1.   A   B   C   D

   (A) management            (C) modern inspection methods
   (B) automation            (D) the shop workers

2. In Dr. Joseph Juran's quality philosophy, it is critical to have participation by          2.   A   B   C   D

   (A) marketing staff         (C) accounting staff
   (B) all parts of the company   (D) none of the above

3. Juran insists that quality goals be          3.   A   B   C   D

   (A) hard to reach          (C) developed from the bottom up
   (B) specific               (D) reviewed quarterly

4. Which of the following is NOT one of the major steps in a sound quality control process?          4.   A   B   C   D

   (A) specification of the goal      (C) production to meet specifications
   (B) inspection of parts produced   (D) meeting productivity goals

5. In the customer's eyes, a company's performance is related to all of the following factors EXCEPT          5.   A   B   C   D

   (A) product quality        (C) delivery time
   (B) product cost           (D) company profits

6. A successful quality control program          6.   A   B   C   D

   (A) increases legal liabilities    (C) reduces warranty repairs
   (B) raises short-term costs       (D) raises long-term costs

Test #55 Statistical Process Control (Section 17, Unit 57)

7.  It is best to catch defects　　　　　　　　　　　　　　　　7.　A　B　C　D

    (A) during the manufacturing cycle  (C) at the customer's location

    (B) after the manufacturing cycle   (D) during the design phase

8.  Which of the following factors does NOT determine the　　　8.　A　B　C　D
    quality

    (A) product design         (C) shop layout and work flow

    (B) manufacturing method   (D) workforce skill

9.  If raw materials are not of suitable quality, the likely　　9.　A　B　C　D
    result is

    (A) lower costs        (C) a poor product

    (B) faster processing  (D) fewer rejects

10. In the traditional manufacturing process, the quality　　　10.　A　B　C　D
    control function was primarily

    (A) an on-line inspection   (C) performed by the supplier

    (B) an off-line inspection  (D) done by customer service

11. In a progressive manufacturing process, who ensures　　　　11.　A　B　C　D
    that top-quality raw materials and sub-assemblies are
    being supplied to the manufacturing processes?

    (A) the supplier          (C) the machine operator

    (B) quality control staff (D) the receiving department

12. The data collected for quality control purposes should　　 12.　A　B　C　D
    be presented in the form of

    (A) checklists           (C) a linear format

    (B) algebraic equations  (D) statistics

13. W. Edwards Deming, a quality expert, claimed that 85% of a company's problems are directly related to

    (A) the quality control department
    (B) management
    (C) shop-floor workers
    (D) suppliers

14. If SPC measurements start to run closer to the high or low quality limits, the process must be

    (A) badly designed
    (B) scheduled for inspection
    (C) slowed down
    (D) changed quickly

15. Which of the following is NOT one of the functions of SPC?

    (A) build process knowledge
    (B) record date
    (C) accept or reject
    (D) examine statistical differences

16. The kind of control chart to be used depends upon the kind of

    (A) manufacturing process used
    (B) data to be studied
    (C) software available
    (D) inspection equipment

17. When using X-bar and R-charts, a point outside the control limits is a signal of

    (A) a defective part
    (B) a defective process
    (C) process bias
    (D) special cause

18. Which descriptive quality control tool is normally used as a final inspection for lot-size parts?

    (A) histogram
    (B) X-bar chart
    (C) line graph
    (D) R-chart

19. Which property of a histogram defines the variability from the aim?

    (A) shape
    (B) width
    (C) height
    (D) centering

    19.  A    B    C    D

20. In a good SPC program, the costs are usually

    (A) maximized
    (B) predictable
    (C) highly variable
    (D) lower then expected

    20.  A    B    C    D

Total—20 marks

# TEST #56
# STEREOLITHOGRAPHY (SECTION 17, UNIT 58)

Stereolithography is a method of rapidly producing a physical model of a part from a CAD or CAM design. This permits designs to be analyzed and modified quickly, and it reduces time-to-market and product development costs.

Place the correct word(s) in the space(s) at the right-hand side of the page to make the statement complete and true.

ANSWER

1. Stereolithography uses an ultraviolet __?__ beam to scan a single cross section of a CAD design into a liquid resin.

    1. _____

2. The resin is changed from liquid to __?__ in thin, accurate layers; the process is repeated until the __?__-dimensional object is completely formed.

    2. _____
       _____

3. A trial model of a part is called a __?__ ; it allows engineers to __?__ their designs early in the manufacturing cycle and locate any __?__ .

    3. _____
       _____
       _____

4. Before the existence of rapid prototyping and manufacturing (RP&M), small errors were often ignored because of the __?__ involved in making changes.

    4. _____

5. The technology of RP&M permits the making of a better product by optimizing the __?__ of the part.

    5. _____

6. Stereolithography begins with data from a __?__ system.

    6. _____

7. In stereolithography, physical models are built one __?__ at a time.

    7. _____

8. After the final part is removed from the vat, it needs to be __?__ before it can be sanded, plated, and painted.

    8. _____

9. RP&M systems must be used for many hours during any given period of time in order to justify their __?__ .

    9. _____

10. Most systems take about a __?__ to produce each layer; therefore, making a model is not very __?__ .

    10. _____
        _____

11. If a fine surface finish is required, the slice thickness should be __?__ ; this requires __?__ time to produce than rougher finishes.

    11. _____
        _____

Test #56 Stereolithography (Section 17, Unit 58)  177

12. If a model needs to last only a few hours for demonstration purposes, it can be made from relatively __?__ materials.

13. Models should be made of stronger materials when the __?__ of their environment is a prime consideration.

14. The required accuracy of the model depends upon the end __?__ of the product.

15. Stereolithography permits the manufacturer to see if designed parts will fit and __?__ properly.

12. _____

13. _____

14. _____

15. _____

_____

Total—20 marks

# TEST #57
# SUPERABRASIVE TECHNOLOGY (SECTION 17, UNIT 59)

Diamond and cubic boron nitride (CBN) superabrasives can cut and grind the hardest materials known. Their special properties make them good choices for many difficult machining operations. They make it possible for the manufacturer to produce parts that would be impossible to process with conventional machining methods.

Consider whether each statement is true or false, then indicate your choice by circling the proper letter in the right-hand column.

ANSWER

1. The superabrasive CBN is the hardest substance known.     1. T F

2. An abrasive material must be harder than the workpiece it will be used to cut.     2. T F

3. Conventional abrasives usually have higher abrasion resistance than superabrasives.     3. T F

4. One of the advantages of superabrasives is their low compressive strength, which reduces tool breakage.     4. T F

5. Superabrasives generally have high thermal conductivity.     5. T F

6. Aluminum oxide is used for truing and dressing grinding wheels.     6. T F

7. Manufactured diamond is used for grinding cemented carbides and carbide/steel combinations.     7. T F

8. When diamond abrasives are coated, this reduces their bonding strength and their useful life, but reduces cutting friction.     8. T F

9. Diamond is one of the best abrasives for grinding ferrous (iron-containing) materials.     9. T F

10. Microfracturing refers to a way that some CBN abrasives resharpen themselves.     10. T F

11. Superabrasive tools usually have high initial costs.     11. T F

12. Because they are used to cut extremely hard materials, superabrasives generally have lower material-removal rates.     12. T F

13. Superabrasives wear out slowly and thus produce parts at a lower cost per piece than conventional cutting tools.     13. T F

14. The first applications of diamond and CBN superabrasives were in grinding wheels.     14. T F

15. Although CBN is a very hard abrasive, it is not able to withstand the high temperatures of production grinding very well.   15. T  F

16. Truing is the process of making a grinding wheel round and concentric with its spindle axis.   16. T  F

17. Dressing is the process of removing some of the bond material from the surface of a grinding wheel.   17. T  F

18. Polycrystalline cubic boron nitride (PCBN) tools have longer lives than cemented carbide and ceramic tools.   18. T  F

19. Superabrasive cutting tools should be used with machines and toolholders that are rigid.   19. T  F

20. Superabrasive cutting tools should be changed (replaced) at the first sign of dullness.   20. T  F

---

Total—20 marks

# TEST #58
# THE WORLD OF MANUFACTURING (SECTION 17, UNIT 60)

During the twentieth century, manufacturing processes changed from mass production and specialization to mass customization and rapid-response systems. Agile manufacturing systems enable companies to adapt and respond quickly to customer needs.

Place the correct word(s) in the space(s) at the right-hand side of the page to make the statement complete and true.

ANSWER

1. Specialized and single-purpose machines were developed between 1900 and 1960 for the __?__ production of identical parts.

1. _____

2. In the early 1900s, many products required __?__ because of unreliable machine tools and human __?__.

2. _____
   _____

3. In the 1990s, the number of machine tools required to produce a given number of products continued to __?__.

3. _____

4. The goal of __?__ manufacturing is to link suppliers, manufacturers, and customers into a superefficient confederation.

4. _____

5. Successful manufacturing in the 21st century will require continuous __?__ and __?__ response.

5. _____
   _____

6. One component of continuous corporate renewal involves the constant upgrading of workforce __?__.

6. _____

7. For a company to respond rapidly to a manufacturing challenge, it must be adaptable and know its own __?__.

7. _____

8. In order for companies to work together successfully, their relationship must be based on __?__; sharing of confidential information may become an __?__ challenge.

8. _____
   _____

9. Rapid response can only be achieved through a __?__ of information technologies and the knowledge and services that allow a company to satisfy the __?__.

9. _____
   _____

10. The key to the agility concept is the rapid collection of __?__.

10. _____

11. In an automatic manufacturing system, accurate measurements can be made whether the parts are __?__ or moving at a __?__ rate of speed.

11. _____
_____

12. __?__ must be designed into the product, not built into it later with inspection or rework.

12. _____

13. Quality is free; what costs money is not doing the job __?__ the __?__ time.

13. _____
_____

14. In the new manufacturing environment, each worker must be a __?__ player.

14. _____

_____
Total—20 marks

# REVIEW TEST #10
# MANUFACTURING TECHNOLOGIES (SECOND HALF)
# (SECTION 17, UNITS 55 TO 60)

**PART 1**

Consider whether each statement is true or false, then indicate your choice by circling the proper letter in the right-hand column.

ANSWER

1. Laser operations rely on photons that cause the release of other photons, thus forming a chain reaction.     1. T F

2. The excimer laser's biggest advantage is its ability to produce high-quality edges on parts with almost no microcracking or thermal damage.     2. T F

3. Rather than specializing in one or two jobs, the ideal industrial robot should be able to perform many different tasks and operations.     3. T F

4. Industrial robots are often equipped with manipulator arms that can move in six or more axes.     4. T F

5. Statistical process control is used to evaluate the stability and predictability of a manufacturing process.     5. T F

6. A recommended way to improve product quality is to have the customer provide input to the product-design process.     6. T F

7. Stereolithography can quickly and accurately produce two-dimensional models of a part.     7. T F

8. Stereolithography enables engineers to test their designs and detect flaws much later in the manufacturing cycle than was previously possible.     8. T F

9. Superabrasives can cut and grind the hardest materials known.     9. T F

10. Although superabrasives can cut very hard materials, they do not maintain their sharp cutting edges as long as conventional abrasives do.     10. T F

11. Because of the great increase in specialization, it takes more machines to produce a limited number of finished products than it did in the early 1900s.     11. T F

12. The philosophy of agile manufacturing is based on treating all customers the same.     12.   T   F

## PART 2

Place the correct word(s) in the space(s) at the right-hand side of the page to make the statement complete and true.

ANSWER

13. The fastest-growing class of lasers are __?__ lasers; they emit ultraviolet (UV) light.     13. _____

14. Liquid lasers are best suited for chemical __?__ .     14. _____

15. __?__ , a relatively new application of lasers, is used to machine cavities in work that is too hard to mill or that cannot be EDM'd.     15. _____

16. The first American industrial __?__ was installed in the early 1960s in an automotive die-casting department; today, they are routinely used to do welding and painting jobs.     16. _____

17. In robotic vision systems, the video system relies on one of two types of light sources, __?__ or __?__ .     17. _____  _____

18. A __?__ is a statistical process control tool that indicates whether a process is holding tolerance and staying within established limits.     18. _____

19. Dr. Joseph Juran, a quality expert, insists that quality goals must be __?__ ; a general goal is not acceptable.     19. _____

20. When using X-bar and R-charts to track process data, a point that lies outside the control __?__ is a signal of special __?__ .     20. _____  _____

21. Stereolithography uses an ultraviolet __?__ beam that changes the resin from liquid to __?__ in thin, accurate layers.     21. _____  _____

22. In RP&M technology, the letter "P" stands for __?__ .     22. _____

23. Important properties of superabrasives include high __?__ resistance and high __?__ conductivity.     23. _____  _____

# PART 3

Select the correct answer for each question and indicate your choice by circling the correct letter in the right-hand column.

ANSWER

24. A very useful tool for measurement applications and image-recognition systems is the
    (A) stereolithography apparatus
    (B) superabrasive cutter
    (C) ultrasonic sensor
    (D) laser

    24.  A  B  C  D

25. An industrial robot is a single-arm device that performs tasks by using its pair of
    (A) grippers
    (B) sensors
    (C) computers
    (D) lights

    25.  A  B  C  D

26. One of the major tools used in modern process control is
    (A) end effectuators
    (B) a prototype
    (C) statistics
    (D) a chemical laser

    26.  A  B  C  D

27. RP&M technology produces a better-quality product because design changes are not so
    (A) frequent
    (B) costly
    (C) necessary
    (D) easy

    27.  A  B  C  D

28. The superabrasive that should be used to grind hard, ferrous metals is
    (A) polycrystalline diamond
    (B) cubic boron nitride(CBN)
    (C) diamond
    (D) silicon oxide

    28.  A  B  C  D

29. Rapid response, which involves sharing, teaming, and cooperating with other companies, can only succeed if it is based on
    (A) a business plan
    (B) cost savings
    (C) statistical analysis
    (D) trust

    29.  A  B  C  D

Total—33 marks